Spaces of Environmental Justice

Antipode Book Series

General Editor: Dr Rachel Pain, Reader in the Department of Geography, Durham University, UK.

Like its parent journal, the Antipode Book Series reflects distinctive new developments in radical geography. It publishes books in a variety of formats – from reference books to works of broad explication to titles that develop and extend the scholarly research base – but the commitment is always the same: to contribute to the praxis of a new and more just society.

Published

Spaces of Environmental Justice
Edited by Ryan Holifield, Michael Porter and Gordon Walker

The Point is to Change it: Geographies of Hope and Survival in an Age of Crisis
Edited by Noel Castree, Paul Chatterton, Nik Heynen, Wendy Larner and Melissa W. Wright

Practising Public Scholarship: Experiences and Possibilities Beyond the Academy
Edited by Katharyne Mitchell

Grounding Globalization: Labour in the Age of Insecurity
Edward Webster, Rob Lambert and Andries Bezuidenhout

Privatization: Property and the Remaking of Nature–Society Relations
Edited by Becky Mansfield

Decolonizing Development: Colonial Power and the Maya
Joel Wainwright

Cities of Whiteness
Wendy S. Shaw

Neoliberalization: States, Networks, Peoples
Edited by Kim England and Kevin Ward

The Dirty Work of Neoliberalism: Cleaners in the Global Economy
Edited by Luis L. M. Aguiar and Andrew Herod

David Harvey: A Critical Reader
Edited by Noel Castree and Derek Gregory

Working the Spaces of Neoliberalism: Activism, Professionalisation and Incorporation
Edited by Nina Laurie and Liz Bondi

Threads of Labour: Garment Industry Supply Chains from the Workers' Perspective
Edited by Angela Hale and Jane Wills

Life's Work: Geographies of Social Reproduction
Edited by Katharyne Mitchell, Sallie A. Marston and Cindi Katz

Redundant Masculinities? Employment Change and White Working Class Youth
Linda McDowell

Spaces of Neoliberalism
Edited by Neil Brenner and Nik Theodore

Space, Place and the New Labour Internationalism
Edited by Peter Waterman and Jane Wills

Forthcoming

Working Places: Property, Nature and the Political Possibilities of Community Land Ownership
Fiona D. Mackenzie

Spaces of Environmental Justice

Edited by

Ryan Holifield, Michael Porter and Gordon Walker

A John Wiley & Sons, Ltd., Publication

This edition first published 2010
Originally published as Volume 41, issue 4 of *Antipode*
Chapters © 2010 The Authors
Book compilation © 2010 Editorial Board of Antipode and Blackwell Publishing Ltd

Blackwell Publishing was acquired by John Wiley & Sons in February 2007. Blackwell's publishing program has been merged with Wiley's global Scientific, Technical, and Medical business to form Wiley-Blackwell.

Registered Office
John Wiley & Sons Ltd, The Atrium, Southern Gate, Chichester, West Sussex, PO19 8SQ, United Kingdom

Editorial Offices
350 Main Street, Malden, MA 02148-5020, USA
9600 Garsington Road, Oxford, OX4 2DQ, UK
The Atrium, Southern Gate, Chichester, West Sussex, PO19 8SQ, UK

For details of our global editorial offices, for customer services, and for information about how to apply for permission to reuse the copyright material in this book please see our website at www.wiley.com/wiley-blackwell.

The right of Ryan Holifield, Michael Porter and Gordon Walker to be identified as the authors of the editorial material in this work has been asserted in accordance with the Copyright, Designs and Patents Act 1988.

Library of Congress Cataloging-in-Publication Data

Spaces of environmental justice / edited by Ryan Holifield, Michael Porter and Gordon Walker.
       p. cm. – (Antipode book series)
  Includes bibliographical references and index.
  ISBN 978-1-4443-3245-2 (pbk. : alk. paper)
  1. Environmental justice. I. Holifield, Ryan. II. Porter, Michael. III. Walker, Gordon.
  GE220.S63 2010
  363.7–dc22

2009053131

A catalogue record for this book is available from the British Library.

Set in 11pt Times by Aptara

01 2010

Contents

Notes on Contributors

Julian Agyeman is Associate Professor and Chair of Urban and Environmental Policy and Planning at Tufts University, Boston-Medford. He is co-founder and co-editor of *Local Environment: The International Journal of Justice and Sustainability*, and his books include *Local Environmental Policies and Strategies* (Longman, 1994), *Just Sustainabilities: Development in an Unequal World* (MIT Press, 2003), *Sustainable Communities and the Challenge of Environmental Justice* (NYU Press, 2005), and *The New Countryside? Ethnicity, Nation and Exclusion in Contemporary Rural Britain* (Policy Press, 2006).

Karen Bickerstaff is a lecturer in Geography at Durham University. Her research is located in the field of environmental social sciences, and increasingly at the interface of geography with science and technology studies. Much of this work has taken a socio-cultural approach to investigating the situated and contextual nature of citizens' experiences of technological and environmental risks. This is linked to an interest in the relationship between space, place, and the construction of risk. She is currently involved in collaborative research (funding under the UK ESRC) concerned with rethinking waste in terms of transformations and flows of materials.

Susan Buckingham-Hatfield is a Senior Lecturer in Geography at Brunel University. Her research interests are primarily in gender/environment relationships, feminist analysis of women, and training and activist research in the academy. She has published a number of books in these areas, including *Gender and Environment* (Routledge, 2000) and *Understanding Environmental Issues* (with Mike Turner; Sage, 2008). Susan combines activism with her own research and teaching, and is on the Board of the Women's Environmental Network. She also broadcasts widely on environmental debates.

Mary Cadenasso is an Assistant Professor in the Plant Sciences at the University of California, Davis, whose research focuses on understanding the ecological and social drivers of and responders to land cover in urban and urbanizing landscapes.

Lisa M. Campbell is the Rachel Carson Assistant Professor of Marine Affairs and Policy, Nicholas School of Environment and Earth Sciences, Duke University. Her work is broadly situated at the intersection of environment and development in rural areas of Latin American, the Caribbean, Southern Africa, and North Carolina. It is informed primarily by political ecology and science studies. She has focused on conservation of endangered species, and specifically of sea turtles, and how conservation conflicts with or enhances local community development. Her most recent research projects examine citizens working for wildlife conservation and their engagements with institutions of science, and local responses to tourism and amenity migration in coastal fishing communities in rural North Carolina.

Trina Filan is a PhD candidate in the Geography Graduate Group at the University of California, Davis. Her dissertation research will focus on whether and how an influx of women farm owner/operators into California agriculture affects individual agricultural practice, gendered professional and personal identities, regional agricultural discourses, and producer–consumer networks. She also is involved in climate change research at the California Energy Commission and in environmental justice research in partnership with UC Davis scholars and graduate students.

Gerardo Gambirazzio is a PhD candidate in the Geography Graduate Group at the University of California, Davis.

Ryan Holifield is an assistant professor in the Department of Geography at the University of Wisconsin – Milwaukee. His research investigates dimensions of environmental justice and injustice in the process of Superfund hazardous waste site remediation and risk assessment. He is also developing a project exploring the nexus between social justice and urban sustainability in Milwaukee.

Rakibe Kulcur is an experienced environmental, health and safety consultant with Enhesa. Since October 2006, she has been doing a PhD at Brunel University, West London. Her PhD focuses on gender structures of Environmental Non-Governmental Organizations in the United Kingdom and Turkey. She has prepared country profiles, audit profiles, regulatory registers, and monitoring reports for Turkey, Azerbaijan, and some European Union countries, and provides rapid-response regulatory research and analysis support to various multinational companies. She studied Public Administration in Izmir, Turkey, and Economics in Göttingen, Germany.

Hilda E. Kurtz is an Associate Professor of Geography at the University of Georgia. Her research explores various influences shaping the character and trajectory of environmental justice activism. To this end, she has theorized grassroots environmental justice organizing as a politics of scale, explored the ways in which multiple ideals of citizenship inform the strategies and goals of environmental justice activists, evaluated the iconography with which environmental justice organizations identify their agendas, and investigated the influence of gender on grassroots environmental justice politics.

Jonathan London is a rural sociologist and environmental planner and is the Director of the UC Davis Center for the Study of Regional Change.

Zoë A. Meletis is Assistant Professor of Outdoor Recreation and Tourism Management at the University of Northern British Columbia. Her academic interests include tourism (especially ecotourism and marine tourism); the relationship between community, environment, and development; speculation and development pressures and local responses to them in attractive locations (for example, coasts and mountains); consumption; and solid waste management, and creative re-use. Her current research projects include an evaluation of a first effort at "greening" The International Sea Turtle Symposium (January 2008), contributing to a project about local perceptions of amenity migrants, coastal gentrification and other development pressures in Down East, NC, and a collaborative paper on the choice of the 2010 Olympic mascots.

Michael Porter is a doctoral candidate in the Department of Earth and Environmental Sciences at the Graduate Center of the City University of New York. His research explores the discourses of brownfields and environmental justice in the context of urban socio-spatial change.

Fraser Shilling is an environmental scientist, with experience in the environmental justice implications of fish contamination, watershed assessment, and the use of geographic information systems (GIS) to provide decision-support for transportation system, pollution management, conservation, and watershed planning.

Julie Sze is an Assistant Professor of American Studies and Director of the Environmental Justice Project of the John Muir Institute of the Environment. Her book, *Noxious New York: The Racial Politics of Urban Health and Environmental Justice* (MIT Press, 2006) examines environmental justice activism in New York City, asthma politics, garbage, and energy policy in the age of privatization and deregulation.

Petra Tschakert is an Assistant Professor in Pennsylvania State University's Department of Geography. Her research activities and practice focus broadly on human–environment interactions and more specifically on environmental change, development, sustainability, knowledge, inequality, and marginalization. Her main interest lies in the theoretical and empirical intersections of political ecology, environmental justice, complex systems science, and participatory research.

Gordon Walker is Chair of Environment, Risk and Justice in the Department of Geography at Lancaster University. He has developed research on environmental justice in the UK, defining an emerging research agenda and influencing its growing profile in policy. This work has included undertaking projects for the Environment Agency, Scottish Executive, Friends of the Earth, and Sustainable Development Research Network, jointly editing journal special issues (in *Local Environment* and *Geoforum*) and jointly convening an ESRC/NERC transdisciplinary seminar series, *Addressing Environmental Inequalities*.

Introduction:
Spaces of Environmental
Justice—Frameworks for Critical
Engagement

Ryan Holifield, Michael Porter
and Gordon Walker

Introduction

It is 13 years since the journal *Antipode* published a collection of papers on environmental justice (1996, volume 28, number 2). At the time the primary academic debates on environmental justice revolved around issues of measurement: is there empirical evidence supporting claims of distributional inequality and disproportionate exposure to environmental risk; is exposure a function of race or class; have poor populations moved to areas with a high concentration of toxic industry; or have toxic industries moved to areas with a high concentration of poor or minority residents? Each of the papers in the 1996 *Antipode* special issue examined these, at the time, exclusively American preoccupations through a refreshingly critical lens. In various ways they investigated how the discourse of environmental justice, along with the rationales, tools and techniques that this discourse enrolled, selectively and problematically represented the issues of concern. At stake was not only the question of whether or not environmental justice existed, but also a critical examination of what the term meant and how institutionalized understandings had impacts upon the practices of activism, research and regulation.

Our aim in 2009 in putting together another collection of critical environmental justice scholarship is to mark how far this has now developed and also to look ahead to how future research might evolve. As we will discuss further, the field has in many ways moved substantially beyond the particular preoccupations and methodologies of the mid-1990s. The set of essays that we have included have their origins at the Association of American Geographers annual conference in Chicago in 2007, in a session specifically seeking critical, theoretically informed contributions. Our intentions were both to bring together various examples of recent critical environmental justice scholarship, and to explore how new insights might emerge by connecting the grounded normative concerns of environmental justice activism to varieties of current social science theory. Given the nature of the conference, the papers also had much to examine and to demonstrate about space and the spatial. The fascination and challenge here was for contributors to think through how space, place, and scale matter to the cases and contexts they examine, and how new frameworks for critical analysis reconfigure those spatialities.

We will have more to say about the particular contributions and insights of each of the chapters in this volume, but first we explore some of the ways in which environmental justice scholarship has recently evolved. Our review and discussion does not attempt to represent the field as a whole, but instead concentrates on the relatively small but growing body of research that attempts to interpret environmental justice activism in the context of critical theory. We contend that the elaboration of two major themes—which themselves overlap and intersect in multiple ways—characterize the development of environmental justice research since 1996: first, more complex and sophisticated conceptualizations of the generation of spaces of environmental inequality; second, ever-increasing scrutiny and analysis of the meanings of *environmental justice* as a discursive frame for activism, policy, and research. These two themes and their intersections also constitute the core concerns of the contributions to this volume.

The Recent Evolution of Environmental Justice Research

There is every indication that environmental justice research is flourishing. Based on a search of the Scopus database since 1996 the number of peer-reviewed articles published with environmental justice in the title or abstract has increased in every year. During the past 5 years alone, there were 425 journal papers published on environmental justice in the social sciences, and no fewer than seven special journal issues dedicated to the topic (*Environmental Policy and Law* 2007; *Geoforum* 2006; *Health and Place* 2007; *Local Environment* 2005, 2008; *Review of European Community and International Environmental Law* 2007; *Society and Natural Resources* 2008).

Much of this body of work continues to follow established quantitative methodologies, which have become more sophisticated and varied in their scope (Beve, Brent and Picou 2007; Buzzelli 2007; Downey 2003; Maantay, Maroko and Herrmann 2007; Mennis and Jordan 2005). However, quantitative research no longer entirely dominates academic discourse on the topic. In contrast to the scholars of the mid-1990s, many of today's environmental justice researchers are situating their work with respect to far broader cross-disciplinary debates about knowledge, representation and meaning, engaging more substantially with explanatory social theory and utilizing a wider diversity of methodologies in investigating the material and political content of socio-environmental concerns. The field in this sense has opened itself up to epistemological and ontological possibilities which were beyond the tightly defined scope of the academic-activist melding that characterized the work of early US environmental justice research. The field has also begun to find productive engagements with other academic traditions, each of which has stimulated the asking of new questions and the exploration of new analytical and explanatory strategies. Wrapped up with this turn towards theoretical and methodological diversity have been two other significant shifts—an enriched understanding of the causes of environmental injustice in its various forms, and an increasingly global scope and perspective, which has revealed the diversity and place specificity of definitions and articulations of environmental justice.

The Production of Spaces of Inequality

One of the key questions driving early research on environmental injustice in the USA, built upon the assumptions of neo-classical economic theory, was whether polluting facilities or land uses were disproportionately sited in communities of color, or whether their spatial allocation simply reflected the dynamics of real estate markets (eg Been 1994; Pastor, Sadd and Hipp 2001). Contributors to *Antipode*'s earlier special issue called for environmental justice research to move beyond this debate, laying out a critical theoretical approach that would begin to situate the production of inequalities with respect to broader social structures and political-economic processes (Heiman 1996; Lake 1996; Pulido 1996, 2000). Recent research has continued and developed this critical commitment in a number of ways. One of the most significant of these has been the emergence of Marxist urban political ecology as a framework for theorizing and analyzing capitalism and class as primary drivers of socio-environmental change (Swyngedouw and Heynen 2003). Although still a relatively new research program, urban political ecology—in both its Marxist and other critical forms—is generating an important body of research, which collectively is deepening our understanding of the production of environmental inequalities by the forces of global capitalism (Heynen 2003; Heynen, Perkins and Roy 2006; Perkins, Heynen and Wilson 2004).

Another important new direction that has begun to emerge is a deeper, more theoretically sophisticated investigation of racism and racialization in the production of environmental inequalities. Deploying

critical social theories of race and racism—particularly Omi and
Winant's (1994) theory of racial formation—and building on Pulido's
(1996, 2000) seminal work, this research is moving far beyond once-
prevalent interpretations of environmental inequalities as the product
of isolated and intentional discriminatory acts. In contrast, it theorizes
environmental injustices as inseparable from racial projects and the
attitudes and beliefs they institutionalize (Park and Pellow 2004; Pellow
2006; Teelucksingh 2007). Nonetheless, as Kurtz (this volume) argues,
there is far more work to be done in order to develop an adequate
understanding of how racism and racialization work to constitute
environmental injustices—just as Buckingham and Kulcur (this volume)
contend that environmental justice research demands greater attention
to the dynamics of gender.

Environmental justice scholarship has also begun to investigate
spaces of environmental injustice generated by the unique historical-
geographical dynamics of colonialism and the oppression of indigenous
populations. In the USA, for instance, we are beginning to see
more environmental justice research address the complexities of tribal
territorial sovereignty and the often conflictual relations among tribes,
states, and the federal government. Ranco and Suagee (2007), for
example, highlight the paradox of "measured separatism" facing
American Indian tribes, which have long struggled to maintain self-
determination and cultural difference in the face of a legal structure
that has historically undermined both. In the late 1980s, Congress
gave the Environmental Protection Agency authorization to treat tribes
"in the manner of states" (TAS) with respect to major environmental
laws, which in principle would finally enable tribes to establish and
enforce environmental standards stringent enough to protect their own
distinctive cultural traditions. However, in practice tribes have faced
two major hurdles—inadequate resources for environmental programs,
and legal decisions that severely limit their power to regulate the
activities of non-Indians within tribal territory—which have prevented
them from overcoming what Ranco and Suagee call "structurally
disproportionate impacts" of environmentally damaging activities. The
distinctive "internal colonialism" of US history has created structural
conditions which have pressured some tribes to accept, for example, the
construction of nuclear waste sites as a viable economic development
strategy (Ishiyama 2003) and disproportionately subjected others to the
adverse effects of US military testing and development (Hooks and
Smith 2004).

Meanwhile, in the US and Canadian Arctic, indigenous populations
disproportionately suffer the environmental consequences of global
warming, generated largely by emissions of greenhouse gases in
distant industrialized and urbanized places (Trainor et al 2007). In

both countries—but particularly the US—these communities have woefully inadequate power to influence decisions that might mitigate the environmental inequalities that affect them. Cases like these have led to the increasing recognition among scholars that environmental injustices refer not only to distributions of environmental hazards and amenities within bounded localities, but also to the ways in which place-specific policies and practices can have consequences that cross national boundaries, affect multiple scales, and extend across global networks. Even if intra-local practices or distributions can be conceived as "just" according to one or another set of criteria, they can result in unjust extra-local outcomes. For Martinez-Alier (2003) these tensions are readily apparent in places like Barcelona, where policies to promote urban sustainability simply displace environmental problems and injustices to regional scales. In New York City, community-based mobilizations against solid waste incineration helped rectify environmental health disparities at the neighborhood scale, but at the same time helped promote regional injustices through the export of the city's waste to distant, often rural localities (Gandy 2002). Heynen (2003) similarly notes that expanding urban forests in wealthier communities might exacerbate local-scale injustices, but can have beneficial impacts on the global environment that contribute to justice at broader scales. This heightened awareness of the scalar dynamics of environmental justice and injustice leads Wolch (2007:379) to advocate for a notion of an urban ecological citizen "whose rights include environmental justice but whose duties and obligations are defined by their ecological footprint": a footprint which extends far beyond the narrowly defined "local". In short, scholars have begun to move well beyond earlier environmental justice research by recognizing that the forces generating spaces of environmental inequality and injustice are far more historically and geographically complex than they once appeared.

Spaces of Meaning: Globalizing and Contextualizing Environmental Justice

One of the characteristics of much early quantitative research was the effort to settle on a standard definition of the term *environmental justice* (along with related terms like environmental equity or environmental racism) that could be "operationalized" and thus subject to measurement (Holifield 2001; Phillips and Sexton 1999). In contrast, recent critical research takes the term's multiple, shifting meanings as an important entry point for inquiry and theorizing. In major recent contributions, Schlosberg (2004, 2007) reconnects the diverse articulations and incarnations of the term by interpreting activist discourses in light of philosophical theories of justice. Building on the work of Young (1990) and Fraser (2000), among others, he argues for a trivalent understanding of environmental justice combining notions of distribution, participation, and recognition. Schlosberg finds that activists, rather than defining environmental justice according to a rigid scheme, combine these elements differently in response to place-specific concerns. Considering activist discourses of environmental justice alongside philosophical theories of justice is transforming our understanding of both. For instance, based on claims advanced by environmental justice activists, Holland (2008) argues that environmental quality should be included as an instrumental value in Nussbaum's "capabilities approach" to justice.

Other research has continued to explore the nuances of the particular ways in which activists articulate and deploy environmental justice discursive frames (Benford 2005). A recurring theme in this research is the diversity of ways that grassroots activists connect issues of environmental injustice with other social concerns. According to Di Chiro (2008), activists' engagement in coalition politics is indicative of a broader ideological commitment to dissolving the boundaries that separate the "environment" from other social issues. These interconnections are readily evident in the way that activists, for instance, seamlessly integrate environmental protests with issues of race (Park and Pellow 2004; Teelucksingh 2007), gender (Buckingham, Reeve and Batchelor 2005; Kurtz 2007), and social reproduction (Di Chiro 2008). Drawing on a Gramscian analysis—and echoing earlier arguments by Harvey (1996)—Kebede (2005) argues that although environmental justice activists emphasize "first order quality of life issues", they also offer a holistic social critique that could form the basis of an alternative hegemonic order.

Perhaps the most dramatic shift in the environmental justice discursive frame has been the geographic dispersion of the term. The discourse built around the term *environmental justice* has its origins in the USA, and in 1996 the vast majority of environmental justice research was produced in and specific to the USA. Since that time mobilizations framed in terms of environmental justice have emerged in many other parts of the world and discourses have diffused and recontextualized. Accordingly, while the corpus of US-based scholarship continues to grow, it is increasingly augmented by studies that examine the relationships among power, inequalities, and environments around the globe—relationships that activists and academics are increasingly identifying in terms of environmental justice and injustice. For instance, recent papers investigate environmental justice issues and activism in Taiwan (Fan 2006; Lyons 2008), Nigeria (Ikporukpo 2004; Omeje 2005), the Middle East (Alleson and Schoenfeld 2007), and Mexico (Grineski and Collins 2008)—to name just a few. In the past 5 years there have been special issues of journals examining environmental justice issues in Canada (*Local Environment* 2008) and the UK (*Local Environment* 2005), as well as edited volumes focusing on Latin America (Caruthers 2008) and South Africa (McDonald 2002).

In the light of this research, the particular geographic biases of past work have become abundantly clear. It is evident that issues of environmental justice, rather than adhering to a particular US-specific definition, manifest differently in different spatial and social contexts. In these new spaces environmental justice discourses have revealed a diversity of forms of environmental good and bad interacting with various forms of social difference (no longer only a matter of race and class) working across and between multiple scales. Throughout the emerging literature on global environmental justice there is a persistent challenge to the tendency in US-based research to focus exclusively on racial minorities and on a small set of environmental hazards (eg Debbané and Keil 2004; Williams and Mawdsley 2006). Whether or not this characterization accurately reflects the diversity of US-based scholarship, it nonetheless clearly points to the importance of place in constructing the meaning of the term *environmental justice* (Schlosberg 2004; Walker and Bulkeley 2006). As Williams and Mawdley (2006) argue, this attention to local context has become particularly important as international organizations have begun creating policies to address existing environmental injustices. No matter how well intentioned these policies may be, if they lack sensitivity to the place specificity of the term's meanings they risk imposing US-based solutions that are ill-suited to social and environmental conditions elsewhere.

While place-specific articulations of diverse injustices often serve as a resource for movement building—whether in the USA or elsewhere—they can also complicate activist struggles. As environmental justice activists build bridges between the environment and a growing array of social concerns, they increase the risk that the term will lose its meaning and the movement will lose its focus (Pellow and Brulle 2005; Sze and London 2008). Benford (2005), for instance, argues that the US movement is weakened not only by the wide range of issues it brings under the umbrella of environmental justice (he identifies over 50 issues on the websites of US environmental justice organizations), but also by its refusal to take the logical step toward a more radical, revolutionary social critique. Meanwhile, relationships between environmental justice activists and other interests may actually blunt the potential for more far-reaching critiques. Macias (2008), for example, documents how in the name of environmental justice, grassroots activists in northern New Mexico sided with timber companies and opposed mainstream environmental organizations in response to a state law regulating forest use. Studies like these raise important questions about the place specificity of the term's deployment, and recall Harvey's (1996) focus on the tension between the impulse of "militant particularism" and the need for an environmental politics that transcends local concerns.

Another arena in which scholars have begun to explore the shifting meanings of the term is the relationship between environmental justice activists and large-scale government and non-government institutions— a relationship which in 1996 was still in the early stages of formation. Holifield (2004), for example, argues that the Clinton administration's approach to environmental justice in US federal government policy ultimately constituted part of an effort to "promote and deepen" the neoliberal project through "neocommunitarian" strategies (Jessop 2002). Within the US Environmental Protection Agency's Superfund program, for instance, implementing environmental justice came in practice to mean little more than finding ways to build trust with community residents, provide vehicles for their participation, and make them aware of federal grant and other financial opportunities. While such institutional initiatives have had some salutary effects, they also threaten to blunt the more radical edges of the environmental justice frame. McCarthy (2004), similarly, argues that large charitable foundations in the USA have created more open and participatory processes and procedures in response to environmental justice activist demands. However, in order to receive funding from these foundations, environmental justice groups have had to embrace less radical objectives and practices. Although the growing institutionalization of environmental justice has given long-overdue recognition to a wide array of environmental and social inequalities, it has raised still more questions about the potential for the environmental justice discursive frame to mobilize a more expansive political project.

Spaces of Environmental Justice

The eight contributions in this volume take up the themes we have identified in the discussion of recent literature and develop them in ways we believe can form the bases for new critical research agendas. Although each of the chapters explores both theoretical and empirical spaces for the study of environmental justice, we have divided the volume into two parts reflecting the chapters' primary emphases. In the first part (Walker, Holifield, Buckingham and Kulcur, and Kurtz), the primary focus is the question of how to conceptualize and theorize issues of environmental justice—and in particular, the forces that generate, stabilize, or even naturalize spaces of inequality and injustice. The chapters in this part highlight four dimensions of environmental injustice in need of deeper theoretical engagement—spatiality, the nonhuman, gender, and the state—and make the case for a diverse array of approaches best fit for such an engagement. While in many ways building on the theoretical interventions of the first part, the second part (Tschakert, Meletis and Campbell, Bickerstaff and Agyeman, and Sze et al) emphasizes the empirical, bringing critical conceptualizations of environmental justice to new social and geographical contexts. Although the second set of chapters also investigates the production of environmental inequalities, it foregrounds questions about the framing of issues in terms of environmental justice, and thereby advances the growing body of literature on the term's articulation, meaning, and deployment. In our discussion below, we seek to bring out some of the ways in which these contributions parallel, complement, and speak to each other, in combination pointing toward numerous unexplored possibilities for the next generation of environmental justice scholarship.

Frameworks for Critical Environmental Justice Research
Although questions of environmental injustice have long been recognized as inherently spatial, critical geographers have only begun to scratch the surface of the spatial dynamics involved in the generation of environmental inequalities. A hallmark of the "first generation" of environmental justice research was a focus on the (Cartesian) space of distribution, concerned specifically with the question of which groups lived in closer proximity to toxic pollution. In the opening chapter of this volume, Walker contends that with the broadening and global diffusion of the environmental justice frame, it has become clear that the spatialities of environmental injustice are multiple—and far more complex than most earlier research recognized. First, the spatiality of distributional justice and injustice cannot be reduced to a matter of simple linear distance. In the classic case of pollution, for instance, we now recognize not only that the paths of chemicals in air, water, and soil are spatio-temporally complex, but also that the distribution of vulnerabilities—among bodies, households, and neighborhoods—does not map neatly onto census-defined demographic groups. In addition, as we consider a wider range of environmental hazards and amenities, from vulnerability to flooding or climate change to the creation of new urban green space, critical geographers must take into account even more idiosyncratic spatialities.

Walker also argues that the focus of early environmental justice research on distributive outcomes in easily mapped space obscured the distinctive spatialities of other key dimensions of justice: responsibility, recognition, and participation. The global proliferation of injustice claims has made it clear, for instance, that where an unwanted risk comes from makes a difference. This is evident not only in the export of solid and hazardous waste from wealthier countries or localities to distant and less advantaged places (see also Bickerstaff and Agyeman, and Meletis and Campbell, this volume), but also in the contributions of commuters to poor air quality in impoverished inner cities, or the threats to coastal and island communities caused by carbon emissions from industrialized countries on the other side of the world. When such places and the people who occupy them become stigmatized and branded with negative environmental associations, other spatialities come into play, which Walker identifies as forms of misrecognition (see also Tschakert, this volume). Finally, the procedural dimension of environmental justice demands that we take account of still more geographies, from the restricted flows and networks of power and decision-making to which participation requires access, to the ambiguous and contested constitution of the "affected community". Walker argues that these diverse, multiple spatialities of environmental justice and injustice call for "new, imaginative, methodologically diverse and theoretically pluralized" interventions—a call taken up in various ways by the remainder of the volume.

A second dimension of environmental injustice in need of deeper theoretical engagement—and the focus of Holifield's chapter—is the status and role of the nonhuman in generating or stabilizing environmental inequalities. One characteristic linking the diverse array of grievances that travel under the name of environmental injustice is the inevitable centrality of nonhuman things: water, air, soil, fish, birds, trees, toxic chemicals, solid and hazardous waste, monitoring and sampling instruments, maps, models, parks, and gardens, just to name a few. Marxist urban political ecology has arguably led the way in developing an agenda for analyzing the role of such nonhuman "actants" in the production of environmental inequalities (Swyngedouw and Heynen 2003). However, while urban political ecologists often attribute their attention to the nonhuman to insights from actor-network theory (ANT), many have dismissed the remainder of ANT as insufficiently critical—and thus of little use in the analysis of power relations and injustices. While some have proposed theoretical syntheses or compromises between Marxist approaches and actor-network approaches, other critical or radical scholars have rejected ANT outright.

In his contribution, Holifield intervenes in this debate by making a case for the utility of ANT in critical environmental justice scholarship, while also arguing that its analytical and political usefulness depends on keeping it separate from Marxist urban political ecology. For one thing, Marxist approaches and ANT take different and incompatible approaches to conceptualizing the *social* (Latour 2005). For Marxist analysis the social refers to the realm of the human, members of which might "mobilize" the nonhuman for various ends. In contrast, for ANT the social refers to a heterogeneous movement of assembling, in which nonhuman agency is central—but nonetheless frequently a site of controversy and uncertainty. If the aim of Marxist approaches is to uncover the social (human) relations that lie underneath and produce environmental inequalities, the aim of ANT is to trace the circulating forms and standards that do the work of stabilizing controversies—or, to put it in more conventionally critical terms, of consolidating hegemony. In order to illustrate the contrasts between the two approaches—and the applicability of ANT to critical environmental justice studies— Holifield's chapter explores possible pathways to analyzing a contested risk assessment on an American Indian reservation. Arguably, the translation and circulation of nonhumans through scientific practice play important roles in generating spaces of inequality, but they remain largely unexplored territory in environmental justice research.

A third theoretical dimension of environmental injustice explored in the volume's first half is gender. Although feminist theory has made inroads into critical environmental justice scholarship (eg Stein 2004), Buckingham and Kulcur contend that the gendering of environmental inequalities demands more attention from activists and analysts alike. First, research and activism should attend more closely to the scales of the body and the household, where the different and unequal environmental experiences and vulnerabilities of men and women have been all too easily overlooked. Also in need of greater attention is the gendering of government institutions and non-governmental organizations, which continue to be dominated by men and by masculinist perspectives on environmental priorities. Buckingham and Kulcur argue that, in both cases, the gendered dimensions of the many forces and relations responsible for producing environmental inequality have remained largely neglected.

If critical environmental justice research is to become more sensitive to such gendering, Buckingham and Kulcur suggest that it will also need to make use of a wider range of theoretical approaches. Ecofeminism, in particular, deserves a fresh look as a means for understanding the gendered production of spaces of environmental inequality and injustice. In addition, critical scholars can do far more to explore the production, experience, and contestation of environmental injustices using the feminist theory of intersectionality, which attends to the complex interactions among oppressions along multiple axes of difference (eg Di Chiro 2008). Undoubtedly, if the future of environmental justice scholarship is to be characterized by "new, imaginative, methodologically diverse and theoretically pluralized" interventions, it will demand a deeper engagement with these and other well-established traditions of feminist thought.

In the concluding chapter of the first part, Kurtz highlights a fourth theoretical blindspot of the current generation of critical environmental justice scholarship: the state, and specifically its centrality in projects of racialization. A hallmark of environmental justice research has long been the investigation of how state policies, programs, and laws affect or address environmental inequalities. But Kurtz contends that such research falls short in its failure to engage with the racialized state apparatus itself, as the preeminent site that structures and delimits the very ways environmental injustices can be expressed and addressed. Her chapter draws out three key tensions in environmental justice research and activism—the race versus class debate, the tension between lay and expert forms of knowledge, and the dialectic of particular and abstract knowledge—and makes a forceful case that lurking behind all of these is the racial state.

In order to move beyond description and address these tensions more productively, Kurtz highlights yet another largely overlooked lens for theoretically pluralized interventions in environmental justice studies: critical race theory, specifically the form it takes in Goldberg's (2002) influential recent theorization of the racial state. For instance, one of the longest-running academic and legal debates in environmental justice has been over whether proving racism in decision-making processes with environmental impacts—such as siting noxious facilities—requires evidence of prejudicial intent, or whether it is sufficient to show disproportionate impacts on populations of color. As Kurtz contends, critical race theory shows how even the latter standard, usually interpreted as more sensitive to the institutional nature of racism, confines the analysis of racism to that which can be measured by the state, specifically through the homogenizing racial categories of the census. Critical environmental justice scholarship would thus do well to move beyond documenting disproportionate impacts, toward investigating the "imbrication of race and racialization in the very structure and outlook of the modern liberal state" (Kurtz, this volume). Without such an investigation, our understanding of racism as a producer of environmental injustices will remain incomplete and impoverished.

Spaces for Critical Environmental Justice Research

The chapters in the second half of the volume extend diverse theoretical perspectives to new empirical spaces in which environmental injustices of various kinds have arisen as concerns, placing particular emphasis on the process of discursive framing. The first two chapters in this part complement and build on Walker's chapter, evealing new spatialities that have accompanied the pluralization of environmental injustices.

The second pair speak both to each other and to Holifield's contribution, highlighting both the framing of geographic scale and the heterogeneous human–nonhuman associations at the heart of environmental justice conflicts. Considered together, the chapters in this part not only reflect the globalization of environmental justice as discursive frame, but also illustrate how environmental justice analysis has matured far beyond the mere documenting of inequalities.

Several of the complex, emerging spatial dimensions of environmental justice that Walker highlights are in abundant evidence in Tschakert's case study. Tschakert's research opens up an important but neglected empirical space of social and environmental injustice: the artisanal small-scale gold mining sector in Ghana. The *galamsey* miners of Ghana occupy an unusual position, in that they suffer the brunt of environmental degradation and occupational hazards that to a considerable extent they create themselves. Consequently, they present a more complex case of social and environmental injustice than, for instance, communities who have been unwillingly burdened by large corporations or governmental institutions with toxic contamination.

By extending to this empirical space the idea of a "contact zone", Tschakert's research also creates a new conceptual space of social and environmental justice. It asks: can *galamsey* miners, many of whom knowingly engage in illegal practices, be justifiably conceived as victims of injustices? Drawing on a range of contemporary critical and radical political theories, Tschakert argues that they can indeed, and that the fundamental issue is a dominant discourse that portrays these miners as violent criminals, instead of recognizing them as citizens with rights to participate. The misrecognition of these miners encompasses not only the disrespect and maltreatment of individuals, but also the institutionalized exclusion of illegal miners as a group from land and decision-making processes. In order to help counter this injustice of misrecognition, Tschakert and her collaborators conducted a series of projects designed to foster "participatory parity", using innovative techniques including participatory hazard mapping, body health mapping, and vision mapping to produce a depiction of an "ideal mining site". Through these approaches, which together built a new contact zone of intercultural knowledge and understanding, the research not only challenges conventional solutions to the conflicts over artisanal mining in Ghana, but also generates a counter-narrative to the discursive representation of *galamsey* miners as criminals. At the very least, this counter-narrative opens up the possibility of more humane and effective government interventions. But Tschakert goes beyond identifying socially and environmentally just alternatives for small-scale gold mining, also providing a model for how novel approaches to participatory research can deepen and enrich the contributions that critical/radical geographers and other scholars can bring to struggles for environmental justice.

Although the space of environmental injustice explored in the chapter by Meletis and Campbell—an ecotourism-based village in Costa Rica— is in many ways a world away from that of the artisanal gold miners in Ghana, there are intriguing parallels that further illustrate Walker's arguments about the pluralizing of environmental justice. At first glance ecotourism, built on its "green" reputation, would appear to be an industry in which claims of environmental injustice would be out of place. However, in the village of Tortuguero, Costa Rica, renowned for its successes in sea turtle conservation, a crisis in the management of solid waste has brought new environmental inequalities to light—and has called the innocence of ecotourism into question.

Meletis and Campbell consider the case for extending an environmental justice conceptual frame to the situation in Tortuguero. They argue that the village's waste crisis fits well with many dimensions of conventional situations of environmental injustice, but that it simultaneously challenges others. On the one hand, Tortuguero's situation resembles those of marginalized environmental justice communities in North America, in that its options for economic development are limited, in part by its isolation and remoteness. In addition, local residents bear disproportionate environmental burdens (in this case, managing excess solid waste without adequate infrastructure), while "outsider" interests reap disproportionate benefits. Finally, community participation is limited and challenging, and it would be hard to defend the claim that development has proceeded with full community consent. But if Tortuguero resembles "classic EJ cases" in all of these ways, the authors show how in other important aspects it departs from them. For example, although not as environmentally benign as it might first appear, ecotourism has nonetheless brought benefits to Tortuguero, making it relatively well-off compared to other villages, and the community has not unified in opposition to a common "enemy". For these and other reasons, it is inappropriate to label Tortuguero as "just another case of environmental injustice". As the case study shows, the unique dynamics and requirements of ecotourism as development strategy complicate efforts to make generalizations about the kinds of injustices communities struggle with in a globalizing economy.

In parallel with Holifield's chapter in the first half, Bickerstaff and Agyeman's contribution explores the possibilities of ANT in critical environmental justice studies. Their chapter explores environmental justice politics in north-east England, a region hit hard by the social and environmental consequences of deindustrialization. Although in the UK neither the concept of environmental justice nor the grassroots politics associated with it have taken hold in any way comparable to the

US context, questions of environmental injustice have emerged in some prominent cases—among them a controversy over whether the Teesside and Hartlepool conurbation should play host to a facility for recycling steel and disposing of contaminants from US military "ghost ships".

In their account of the case, Bickerstaff and Agyeman not only expand instructively on Kurtz's (2003) conceptions of scale frames and counter-scale frames in the politics of environmental justice, but also provide further elucidation of the centrality of nonhuman "actants" in environmental justice conflicts. On the one hand, they demonstrate how facility opponents successfully translated the local injustice of polluting the Teesside area into a national and international issue. On the other, they show how the recycling firm ultimately managed to overcome opposition to its ship-dismantling operation by framing opposition to the facility as itself an injustice. This "counter-scale frame" emphasized both the regional scale, in desperate need of economic development, and the global scale, in which the operation would otherwise shift to localities with even more disadvantages and weaker environmental standards. However, Bickerstaff and Agyeman's account goes beyond providing another illustration of the scaling of environmental justice politics. By tracing the roles of nonhuman actors in the competing assemblages—for instance, the divisions created by the physical presence of ships in Hartlepool—the authors challenge the common understanding of framing in environmental justice politics as the work of human actors alone. In doing so, they provide further support for Holifield's claim that critical environmental justice research would do well to conceive of "social construction" not as a distinctively human process, but as the "construction of the social" undertaken by human–nonhuman associations (Latour 2005).

Conflicting scale frames and human–nonhuman associations also play central roles in the final chapter by Sze et al, which investigates the representation of environmental justice in the State of California's management of the Sacramento–San Joaquin Delta. The "Delta Vision" planning process has largely framed the Delta as a resource that provides services at the scale of the State—and one that consequently requires management at the State level. So by framing environmental justice as a narrow, local, and basically irrational "special interest" (cf Kurtz, this volume), rather than a comprehensive framework that should permeate the entire decision-making process, the Delta Vision both ignores state law and marginalizes the needs and voices of the many diverse socially vulnerable populations who live within the Delta region itself. Environmental justice advocates in California have sought to reframe the spatiality of both "environmental justice" and "the Delta", bringing the latter to the fore as a place of everyday life, a dynamic "spatial patchwork" rather than a "placeless space" that exists only to serve a nebulous "common good" defined at the abstract scale of the State. At the same time, they (and the authors) take the Delta Vision to task for taking an approach to water management that ignores the historical geography of the region: in particular, the displacement of Native tribes and the degradation of tribal lands, as well as the exploitation of racialized populations in the massive re-engineering of the Delta to serve industrial agriculture and distant municipalities.

In addition to its contribution to research on the scaling of environmental justice, the chapter by Sze et al offers yet another intervention into debates on the relation of environmental justice to the broader political economy of "socio-natures". Their response to Swyngedouw and Heynen's (2003) critique of environmental justice from the standpoint of Marxist urban political ecology follows a different tack than Holifield's, highlighting the former's characterization of environmental justice politics as reformist and locally focused. The authors cite in particular the work of the Environmental Justice Coalition for Water, a statewide network of grassroots organizations that has emphasized goals and needs that transcend specific localities, promoting the adoption of environmental justice as overarching principle for the State of California. While the authors acknowledge that the Coalition's strategies have in fact been largely reformist rather than radical, they also point to ways in which a critique of the Delta as hybrid socio-natural assemblage might open up possibilities for a more deeply radical politics of environmental justice. To be sure, Sze et al present an analysis that is "critical" in at least two senses: on the one hand, situating the Delta Vision with respect to broader dynamics of state and capital, on the other hand, offering a direct and pointed critique of a flawed planning and policy-making process. They also illustrate the relevance of Kurtz's call for environmental justice scholarship to turn its attention to the state as both stakeholder and preeminent site of racialization.

Future Directions in Unstable Times: Questions for a Critical Environmental Justice Research Agenda

This introductory essay has, we hope, demonstrated that environmental justice research is undergoing something of a rejuvenation. It is finding new spaces to work in, new spatialities to work with, and new ways of engaging with justice and fairness in the environment. Our focus has been on critical, theoretically informed analysis, and the chapters each demonstrate how this has a role to play in scrutinizing, questioning, and sometimes reformulating the nature of what constitutes environmental injustice, the practices of environmental justice activism and the responses and strategies of the state and other actors. There is much scope and potential for further work in this vein, whether building on the essays we have collected together here or pursuing other potential directions and inflections. While encouraging such moves, we must also caution that they are not without their challenges; we accordingly complete our discussion by highlighting three of these.

First is the broader challenge to critical or radical social science, and specifically to radical geography, laid down by Olson and Sayer (2009) in an *Antipode* symposium published earlier this year. Addressing the decline of "normative thinking", they argue that there is both uncertainty and inadequacy in how critique is understood and being practised. The social sciences, they argue, have become increasingly cautious about being normative because of the "dangers of ethnocentric critiques" (2009:182), while radical geographers in particular have tended too easily to adopt partisan positions that implicitly assume a common understanding of what constitutes the good and the bad. The environmental justice scholarship we have reviewed is beginning to some degree to recognize the need for more careful analytical reasoning about the constitution of justice and injustice, in particular as it examines more heterogeneous settings for making claims. However, there is more to be worked through here (see also Walker 2009). As we have discussed, tensions endure regarding the extent to which universal principles or formulations of environmental justice can be identified, or whether particular, situated understandings of the term are necessary to reflect the global diversity of materialities, values and normativities. This is not simply an academic point; to the contrary, it is fundamental to whether and how environmental justice might cohere or fragment as a political project. While for Schlosberg (2004:535) there is virtue in the "salad bowl" in which there is "unity without uniformity", for Sze and London (2008:1347) there is a concern that "if environmental justice can mean almost anything, does it risk a dilution and even loss of meaning and purpose?" From our perspective, the investigation of environmental justice as a contested, complex discursive frame has just begun—and needs to continue.

A second interconnected challenge relates to where and how critical environmental justice scholarship is to be practised, particularly within geography. In his exploration of the spaces of critical geography, Bromley (2007) finds a breadth that extends from the internal spaces of the academy, through to various forms of external engagement with grassroots activism and, for some at least, with policy and state institutions. A recognized need to work beyond the confines of the university has characterized the practice of environmental justice researchers from the earliest stages of political mobilization in the USA, but how this is to be sustained is a key question. A turn towards theorization can simultaneously become a turn towards obfuscation, disconnection from everyday vocabulary, and an increased distance from the material substance of unjust patterns and processes. Similarly a translation of environmental justice discourse into new political spaces can manifest itself as a de-radicalization into the language of management and technical assessment. There is therefore a need for environmental justice scholarship to actively work at its connections to activism and its engagement with those at the sharp end of injustice, however it is understood, and to bring theory to bear in meaningful ways into praxis and diverse forms of public engagement. Tschakert's contribution to this volume provides an excellent model of one way to do this, and we hope it inspires other new methods and approaches to geographic scholarship that is both theoretically grounded and practically engaged.

The third and perhaps biggest challenge centers on the radical and rapid economic, social, political, and environmental change that is currently enveloping much that has been taken for granted around the world. As we write, global capitalism is undergoing one of the greatest crises of its still relatively short history, and global warming is disrupting social and ecological relations in ways that are equally dramatic—if not more so. While the extent and implications of this complex and systemic dynamism are uncertain, unknowable, and rapidly unfolding, we can speculate as to how the concerns of environmental justice activism and scholarship might fare. Will the discourse of justice in the environment become more or less "relevant", the politics more or less possible, and the need for critical engagement more or less pressing?

As might be expected, our speculations move in a variety of sometimes conflicting directions. On the one hand, a reining in of "excess" consumption at the global scale may prove progressive in its environmental implications, perhaps mitigating to some degree the scale of disparities in resource consumption and its unequal environmental consequences. A move away from unbridled neoliberalism towards a greater recognition of the need for regulation and intervention by the state may have welcome consequences for a wide range of social and socio-environmental concerns. In the US, for instance, environmental and social justice activists are expressing optimism about the priorities reflected in the Obama administration's proposed budgets and stimulus packages, which include funds to create new jobs by "greening" the energy infrastructure and restoring degraded environments and ecosystems like the Great Lakes. At the same time, there is new hope that the USA may finally take meaningful action on global climate change—and rising awareness that global warming, itself the product of vast inequalities in production and consumption, is producing new disparities and injustices at multiple scales. Meanwhile, in parts of the world that are now undergoing rapid industrialization and urbanization, the crises of the present moment may offer opportunities to reorient economic development in less socially and environmentally damaging ways.

On the other hand, collapsing incomes and rising unemployment can only exacerbate problems associated with "underconsumption" (food, water, energy) of key environmental resources by disadvantaged social groups, and we face the risk that the push for renewed aggregate economic growth will ultimately sideline social and environmental concerns. For instance, if the unequal access to fuel, food, and water faced by the world's poor—in many places compounded by changing climatic conditions—was already a crisis of global proportions, it is now likely to grow worse. And if state spending becomes narrowly devoted to supporting financial and industrial infrastructures, other public concerns may become increasingly crowded out, relegating unfairness and injustice in the environment to an "unfortunate but necessary" consequence of maintaining capital accumulation in whatever form it is able to take.

Although it is far from clear what implications these developments will ultimately have for the production or rectification of environmental inequalities, it is clear to us that critical geographers and our colleagues in related disciplines must seize the moment and take the lead in addressing them. Whether we see an intensification of environmental inequalities, steps toward realizing "just sustainability" (Agyeman, Bullard and Evans 2002), or (most likely) both, we need to continue developing critical geographic approaches to environmental justice and injustice that can be models of empirical rigor, theoretical sophistication, and practical engagement. To that end, we hope the chapters in this volume will together constitute a significant contribution, and one we hope will help deepen a growing environmental justice literature worthy of being called "critical".

References

Agyeman J, Bullard R and Evans B (2002) Exploring the nexus: Bringing together sustainability, environmental justice and equity. *Space and Polity* 6(1):70–90

Alleson I and Schoenfeld S (2007) Environmental justice and peacebuilding in the Middle East. *Peace Review* 19(3):371–379

Been V (1994) Locally undesirable land uses in minority neighborhoods: Disproportionate siting or market dynamics? *Yale Law Journal* 103:1383–1422

Benford R (2005) The half-life of the environmental justice frame: Innovation, diffusion, and stagnation. In D Pellow and R Brulle (eds) *Power, Justice, and the Environment* (pp 37–54). Cambridge, MA: MIT Press

Beve C A, Brent K M and Picou S J (2007) Environmental justice and toxic exposure: Towards a spatial model of physical health and psychological well-being. *Social Science Research* 36:48–67

Bromley N (2007) The spaces of critical geography. *Progress in Human Geography* 32(2):285–293

Buckingham S, Reeves D and Batchelor A (2005) Wasting women: The environmental justice of including women in municipal waste management. *Local Environment* 10(4):427–444

Buzzelli M (2007) Bourdieu does environmental justice? Probing the linkages between population health and air pollution epidemiology. *Health and Place* 13:3–13

Caruthers D (ed) (2008) *Environmental Justice in Latin America: Problems, Promise, and Practice*. Cambridge, MA: MIT Press

Debbané A M and Keil R (2004) Multiple disconnections: Environmental justice and urban water in Canada and South Africa. *Space and Polity* 8(2):209–225

Di Chiro G (2008) Living environmentalisms: Coalition politics, social reproduction, and environmental justice. *Environmental Politics* 17(2):276–298

Downey L (2003) Spatial measurement, geography and urban racial inequality. *Social Forces* 81:937–954

Environmental Policy and Law (2007) Environmental Justice and Regional and National Matters. 37(2–3)

Fan M F (2006) Environmental justice and nuclear waste conflicts in Taiwan. *Environmental Politics* 15(3):417–434

Fraser N (2000) Rethinking recognition. *New Left Review* 3:107–120

Gandy M (2002) *Concrete and Clay: Reworking Nature in New York City*. Cambridge, MA: MIT Press

Geoforum (2006) Special Issue: Geographies of Environmental Justice. 37(5)

Goldberg D T (2002) *The Racial State*. Oxford: Blackwell

Grineski S E and Collins T W (2008) Exploring patterns of environmental injustice in the Global South: Maquiladoras in Ciudad Juarez, Mexico. *Population and Environment* 29(6):247–270

Harvey D (1996) *Justice, Nature and the Geography of Difference*. Cambridge, MA: Blackwell

Health and Place (2007) Special Issue: Environmental Justice, Population Health, Critical Theory and GIS. 13(1)

Heiman M K (1996) Race, waste, and class: New perspectives on environmental justice. *Antipode* 28:111–121

Heynen N C (2003) The scalar production of injustice within the urban forest. *Antipode* 35(5):980–998

Heynen N, Perkins H A and Roy P (2006) The political ecology of uneven urban green space: The impact of political economy on race and ethnicity in producing environmental inequality in Milwaukee. *Urban Affairs Review* 42(1):3–25

Holifield R (2001) Defining environmental justice and environmental racism. *Urban Geography* 22:78–90

Holifield R (2004) Neoliberalism and environmental justice in the United States Environmental Protection Agency: Translating policy into managerial practice in hazardous waste remediation. *Geoforum* 35(3):285–297

Holland B (2008) Justice and the environment in Nussbaum's "capabilities approach": Why sustainable ecological capacity is a meta-capability. *Political Research Quarterly* 61(2):319–332

Hooks G and Smith C L (2004) The treadmill of destruction: National sacrifice areas and Native Americans. *American Sociological Review* 69(4):558–575

Ikporukpo C O (2004) Petroleum, fiscal federalism and environmental justice in Nigeria. *Space and Polity* 8(3):321–354

Ishiyama N (2003) Environmental justice and American Indian tribal sovereignty: Case study of a land-use conflict in Skull Valley, Utah. *Antipode* 35(1):119–139

Jessop B (2002) Liberalism, neoliberalism and urban governance: A state-theoretical perspective. *Antipode* 34(3):452–472

Kebede A (2005) Grassroots environmental organizations in the United States: A Gramscian analysis. *Sociological Inquiry* 75(1):81–108

Kurtz H E (2003) Scale frames and counter scale frames: Constructing the social grievance of environmental injustice. *Political Geography* 22:887–916

Kurtz H E (2007) Gender and environmental justice in Louisiana: Blurring the boundaries of public and private spheres. *Gender, Place and Culture* 14(4):409–426

Lake R (1996) Volunteers, NIMBYs, and environmental justice: Dilemmas of democratic practice. *Antipode* 28(2):160–174

Latour B (2005) *Reassembling the Social: An Introduction to Actor-Network-Theory.* Oxford and New York: Oxford University Press.

Local Environment (2005) Special Issue: Environmental Justice in the UK. 10(4)

Local Environment (2008) Special Issue: Inequality and Sustainable Consumption. 13(8)

Lyons D (2008) The two-headed dragon: Environmental policy and progress under rising democracy in Taiwan. *East Asia* 26:57–76

Macias T (2008) Conflict over forest resources in Northern New Mexico: Rethinking cultural activism as a strategy for environmental justice. *Social Science Journal* 45(1):61–75

Maantay J, Maroko A and Herrmann C (2007) Mapping population distribution in the urban environment: The cadastral-based expert daysmetric system (CEDS). *Cartography and Geographic Information Systems* 34(2):77–102

Martinez-Alier J (2003) Scale, environmental justice, and unsustainable cities. *Capitalism, Nature, Socialism* 14(4):43–63

McCarthy D (2004) Environmental justice grantmaking: Elites and activists collaborate to transform philanthropy. *Sociological Inquiry* 74(2):250–270

McDonald D (ed) (2002) *Environmental Justice in South Africa.* Columbus, OH: Ohio University Press

Mennis J and Jordan L (2005) The distribution of environmental equity: Exploring spatial nonstationarity in multivariate models of air toxic releases. *Annals of the Association of American Geographers* 95(2):249–268

Olson E and Sayer A (2008) Radical geography and its critical standpoints: Embracing the normative. *Antipode* 41(1):180–198

Omeje K (2005) Oil conflict in Nigeria: Contending issues and perspectives of the local Niger Delta people. *New Political Economy* 10(3):321–334

Omi M and Winant H (1994) *Racial Formation in the United States: From the 1960s to the 1990s.* 2nd ed. New York: Routledge

Park L S H and Pellow D N (2004) Racial formation, environmental racism, and the emergence of Silicon Valley. *Ethnicities* 4(3):403–424

Pastor M, Sadd J and Hipp J (2001) Which came first? Toxic facilities, minority move-in, and environmental justice. *Journal of Urban Affairs* 23:1–21

Pellow D N (2006) Social inequalities and environmental conflict. *Horizontes Antropologicos* 12(25):15–29

Pellow D N and Brulle R (2005) Power, justice, and the environment: Toward critical environmental justice studies. In D N Pellow and R Brulle (eds) *Power, Justice, and the Environment: A Critical Appraisal of the Environmental Justice Movement.* Cambridge, MA: MIT Press

Perkins H A, Heynen N and Wilson J (2004) Inequity in an urban reforestation program: The impact of housing tenure on urban forests. *Cities* 21(4): 291–299

Phillips C V and Sexton K (1999) Science and policy implications of defining environmental justice. *Journal of Exposure Analysis and Environmental Epidemiology* 9(1):9–17

Pulido L (1996) A critical review of the methodology of environmental racism research. *Antipode* 28(2):142–159

Pulido L (2000) Rethinking environmental racism: White privilege and urban development in Southern California. *Annals of the Association of American Geographers* 90(1):12–40

Ranco D and Suagee D (2007) Tribal sovereignty and the problem of difference in environmental regulation: Observations on "measured separatism" in Indian country. *Antipode* 39(4):691–707

Review of European Community and International Environmental Law (2007) Special Issue: Articles on Environmental Rights. 16(3)

Schlosberg D (2004) Reconceiving environmental justice: Global movements and political theories. *Environmental Politics* 13(3):517–540

Schlosberg D (2007) *Defining Environmental Justice: Theories, Movements, and Nature.* New York City, NY: Oxford University Press

Society and Natural Resources (2008) Special Issue: Third World Environmental Justice. 21(7)

Stein R (ed) (2004) *New Perspectives on Environmental Justice: Gender, Sexuality, and Activism.* New Brunswick, NJ: Rutgers University Press

Swyngedouw E and Heynen N C (2003) Urban political ecology, justice and the politics of scale. *Antipode* 35:898–918

Sze J and London J K (2008) Environmental justice at the crossroads. *Sociology Compass* 2:1331–1354

Teelucksingh C (2007) Environmental racialization: Linking racialization to the environmental in Canada. *Local Environment* 12(6):645–661

Trainor S, Chapin F S, Huntington H P, Natcher D C and Kofinas G (2007) Arctic climate impacts: Environmental injustice in Canada and the United States. *Local Environment* 12(6):627–643

Walker G P (2009) Environmental justice and normative thinking. *Antipode.* 41(1):203–205

Walker G P and Bulkeley H (2006) Geographies of environmental justice. *Geoforum* 37(5):655–659

Williams G and Mawdsley E (2006) Postcolonial environmental justice: Government and governance in India. *Geoforum* 37(5):660–670

Wolch J (2007) Green urban worlds. *Annals of the Association of American Geographers* 97(2):373–384

Young I M (1990) *Justice and the Politics of Difference.* Princeton, NJ: Princeton University Press

Part I: Frameworks for Critical Environmental Justice Research

Chapter 1

Beyond Distribution and Proximity: Exploring the Multiple Spatialities of Environmental Justice

Gordon Walker

Introduction

Over the last decade environmental justice has evolved both as a political discursive frame and as a focus of academic study. The material and sociological themes of concern for activists and researchers are now extending far beyond the local distribution of pollution, risk and race to include many other environmental concerns and many other forms of social difference. The spatio-cultural and institutional contexts in which justice claims are being made and justice discourses are being articulated are globalising far beyond the USA to include, for example, South Africa (London 2003), Taiwan (Fan 2006), Australia (Hillman 2006), the UK (Agyeman and Evans 2003), New Zealand (Pearce et al 2006), Sweden (Chaix et al 2006), Israel (Omer and Or 2005), and global contexts (Adeola 2000; Newell 2005). In addition, the established representation of environmental justice as only a matter of socio-spatial maldistribution (Dobson 1998) is being replaced by a conceptualisation that is more open to other notions of justice figuring in the evidence gathering and claim making of environmental justice activists and academic researchers (Schlosberg 2004, 2007; Wenz 1988).

In this chapter I argue that this substantive and theoretical pluralism has important implications for geographical inquiry and analysis, meaning that multiple forms of spatiality are entering our understanding of what it is that makes and sustains environmental injustice in different contexts. In this light the simple geographies and spatial forms evident in much "first-generation" environmental justice research are insufficient and inadequate to the tasks of both revealing inequalities and understanding the processes through which these are (re)produced. Instead a multidimensional understanding of the various ways in which environmental justice and geography are co-constituted is needed. Following Harvey's (1996:5) observation that concerns about justice "intertwine with the question of how to understand foundational geographical concepts", I argue that spatialities of different forms, of different things and working at different scales need to be integral in our understanding of the multiplicity of contemporary environmental justice concerns and claims. I develop this argument by examining a purposefully diverse range of examples of socio-environmental concerns that have been the focus of recent and more established research and political activism. This breadth stretches the spatialities involved from simple local proximities to more complex scaled spatial relations and flows, and brings forward multiple ways in which wellbeing, vulnerability and environment are spatially intertwined. Space, as many others have argued, is constructed by and through social practices, including those of activists and researchers. Given the variety of spatialities available (Leitner, Sheppard and Sziarto 2008) the chapter therefore seeks to identify systematically those that are being deployed within the evolving environmental justice frame and to consider the implications of both the particularities and the diversity that is revealed.

The framework used to structure this analysis draws on justice theory to move through three understandings of what defines the "justice" in environmental justice (Schlosberg 2007). First, distributional understandings of justice in terms of the unequal distribution of impacts, the unequal distribution of responsibilities and the spatialities that are implicated within these. Second justice as recognition (Fraser 1997; Honneth 2001) in terms of the processes of disrespect, insult and degradation that devalue some people and some place identities in comparison to others. Third, justice as participation and procedure (Hunold and Young 1998; Young 1990) in terms of how geography plays into the inclusions and exclusions of environmental decision-making. In using this framework I seek to promote a move beyond the distributional in geographical research towards a fuller and more integrated understanding of what the spatiality of environmental justice can constitute.

I begin the discussion by mapping out in more detail how the scope and meaning of environmental justice has broadened and pluralised over the last decade. This then provides the context for considering the spatiality of environmental justice as revealed within different conceptualisations of justice and across a diversity of environmental justice concerns.

Pluralising Scope and Meaning

The history and origins of environmental justice as a term, a set of ideas and a focus for political activism in the USA are well known and well documented (Bryant 2003; Bullard 1999). The core concern with the burdens of pollution and risk associated with waste and industrial sites, and how these sites were distributed, particularly in relation to race, was distinctive and challenging to conventional environmentalism (Shrader-Frechette 2002) and in many ways particular to time and place. The overwhelming majority of the "first-generation" research literature on environmental justice worked within this frame and conception, documenting the distribution of hazardous sites and racial groups, the historical evolution of these socio-spatial patterns (eg Hurley 1995) and the successes, failures and strategies of place-based environmental justice activism.

While this particular conceptualisation of environmental justice remains influential in the USA (Bullard et al 2007) and has, to some degree, reproduced itself as the environmental justice frame has moved into other countries, over the last decade there has been a broadening of the scope and understanding of what environmental justice constitutes (Walker and Bulkeley 2006). This has both substantive and more theoretically driven dimensions. In substantive terms there has been a broadening of the environmental and social concerns positioned within an environmental justice framing moving beyond only environmental burdens to include environmental benefits and resources in various forms (Laird, Cunningham and Lisinge 2000; Mutz, Gary and Douglas 2002; Schroeder 2000). A review of the content of activist group web sites in the USA identified 50 distinct and varied environmental themes (Benford 2005) and recent writing in the USA has focused on increasingly diverse concerns—including, for example, access to food (Williams 2005), flood disaster (Sze 2006), forest management (Carey 2002) and transport (Targ 2005). A similarly pluralistic and expansive framing of environmental justice also exists in the UK, one of a long list of countries in which the discourse of environmental justice is now appearing. A review of evidence of the relationship between environmental and social justice undertaken in 2004 covered 21 topics encompassing environmental goods (such as greenspace, food and water) as well as bads, and issues of environmental consumption and service provision (Lucas et al 2004). An earlier agenda-setting report produced jointly by a UK research council and Friends of the Earth England and Wales (Stephens, Bullock and Scott 2001:3) also firmly sought to go "beyond the US approach", incorporating international and global environmental concerns, such as climate change and resource extraction, and intergenerational justice issues. The coming together of sustainability and environmental justice discourses, in part through the conceptualisation of "just sustainability" (Agyeman and Evans 2003), has been a significant part of this broadening and globalising process. In parallel, the initial concentration on intentional environmental racism has also shifted to encompass more nuanced understandings of structural racism and intersections between race and class (Pulido 1996, 2000); and attention has been increasingly given to many other forms of social difference. These have included research focused on environmental justice in relation to poverty and deprivation (the dominant concern in the UK; Walker et al 2003), age (Chaix et al 2006), disability (Charles and Thomas 2007) and gender (Buckingham-Hatfield et al 2005; Kurtz 2007).

Alongside this broadening of scope a more developed and richer understanding of the multiple meanings of environmental justice has emerged. In part this has stemmed from better recognition that environmental justice activism has always been concerned with more than questions of distribution (Wenz 1988). While distributional justice—who gets what in the environment—has undoubtedly been the dominant mode of representing the claims of environmental justice activists, particularly in the USA (Schlosberg 2007), there has always been a strong procedural justice dimension to stated environmental justice principles and objectives—the "reclaiming democracy" of Shrader-Frechette (2002). Justice claims have routinely extended beyond the distributional to include matters of fairness in process and regulation, inclusion in decision-making and access to environmental information (Dunion 2003; Hampton 1999; Hunold and Young 1998; Lake 1996; Petts 2005). Wider developments in justice theory have similarly moved beyond the distributional to emphasise the role of process, procedure and recognition in underlying the production of unequal outcomes (Fraser 1997; Young 1990). The work of David Schlosberg (2002, 2004, 2007) has been particularly influential in integrating different theoretical perspectives into a plural understanding of environmental justice and demonstrating how both procedure and recognition[1] are evident components of environment justice discourses (discussed further below).

In these ways environmental justice has become increasingly different to when and where it began. As the objects of attention and our understandings of the ways in which justice claims act as normative evaluations of socio-environmental conditions have diversified, a more intrinsically involved field of study has emerged. For all forms of disciplinary scholarship this has implied the need to rethink or refashion tools of analysis, including those of geography, which has made a significant contribution to the research field. It is therefore to matters of geography and the ways in which the spatial is conceived and entwined within a pluralised understanding of environmental justice that the rest of the chapter now turns.

Geographies of Distribution and Inequality

Distributional notions of justice for a long time dominated justice theory and thinking, and have been central to much engagement within geographical scholarship (Smith 1994). In cities, in rural spaces and in the global economy, distributional inequalities, including those of the environment, have a demonstrable spatial expression and constitution. However, this spatiality is not unproblematic or given. Rather the ways in which environmental inequalities are understood, the nature of the socio-environmental relations that are at issue and the evidence that is used to give credence to claims of injustice gives importance to the spaces of different social and environmental categories and to different notions of space itself.

Space as Proximity

For the early headline claims of the environmental justice movement, spatially articulated socio-environmental inequalities were absolutely central. Studies of the socio-spatial patterning of the locations of waste, landfill and industrial sites and their proximity to populations of different racial make-ups were enormously influential in providing evidence for activists seeking to interconnect, systemise and upscale local protests against the siting of such facilities in black and poor communities. Various reviews have documented the enormous number of GIS-based studies undertaken in different parts of the USA (Bowen 2002; Brown 1995; Bullard et al 2007; Holifield 2001) in which the test of environmental injustice (and for some of environmental racism) was distributional and statistical, seeking evidence of disproportionate bias in locations of particular types of installations towards racial minority populations (Low and Gleeson 1998). As documented by Holifield (2004) for the US Environmental Protection Agency this translated into problematic attempts to statistically codify what constituted in these terms an "environmental justice community". Space was central to these research and policy tasks, conceived in flat, Cartesian terms—straight line proximity, or coincidence of site grid references within census boundaries. People were given a racial and sometimes class identity, and counted and compared in aggregate to establish patterns of over or under-representation in spatial terms.

This simple notion of geography and simple epistemology of inequality proved sufficient and effective when environmental justice remained in its initial narrow conception, and provided for some time, within the geographical research community and beyond, the core of environmental justice scholarship. However, its limitations have had consequences for the obscuring of what are in practice far more involved and multifaceted relations between environmental features and human wellbeing, and the potential hiding of forms of inequality that do not fall into such a simple and particular spatial form.

Remaining, for the moment, within the territory of pollution and technological risk, there are evident limitations in using census boundaries or circles drawn around grid references to estimate who is somehow "at risk" from a waste site or factory producing pollutants. Pathways of pollutants are far more involved than this, leading to exposures and potential impacts that cannot be captured through simple proximity measures (Bowen 2002; Bowen and Wells 2002; Brown 1995; Liu 2001; Zimmerman 1994). More sophisticated analytical techniques which take better account of the more complex, dynamic and fluid spatialities of dispersion as pollutants move, are transformed and received by people located in different spaces and their resulting disease epidemiology have consequently been called for (Buzzeli 2007; O'Neill et al 2003). However, it is not simply a matter of developing better scientific tools that work with more involved notions of the spatial relation between pollution source and person, which can establish the spatiality of "who gets what" in a more sophisticated manner (Pulido 1996). It is also necessary to understand how the body, the household and wider social context are also implicated in the social patterning of impacts on health and wellbeing.

Kuehn (1997), writing from within an environmental justice framing, gives rare attention to the body (although see also Getches and Pellows 2002), arguing that risk assessment practices being widely applied in the USA and beyond were failing to recognise that bodies of different ages, races and genders were sensitive to and harmed by pollutants to different degrees. Locked into applying assessment methodologies to an "average white male reference man" he claims resulted in a "risk assessment characterisation that fits far less than half the nation's population, because the majority are women, children, the elderly, sick or people of colour" (Kuehn 1997:268). This institutionalised bias and lack of attention to the corporeogeographies of pollution and inequality (Longhurst 2001) has become all the more significant as the social differences of environmental justice have extended beyond race to include gender, age and disability. Furthermore while all bodies are not physiologically equal, neither, clearly, is the social context for people in households, living and working within communities with differential access to resources, to healthcare, to healthy and good-quality food and so on. Pollution is socially contextualised, intersecting with life course, class and poverty so that impacts of "equal doses" are not equally experienced or coped with—an observation that extends to the unevenness of the psycho-social as well as the physiological impacts of living with sources of risk (Bickerstaff and Walker 2003; Gee and Payne-Sturges 2004).

While these may seem like obvious points, it is rare to find them made within geographical scholarship on environmental justice (see Cutter 1995 for an early exception). Much of the classic geographical contribution has been so locked into a frame concerned with the spatiality and politics of siting, of discriminatory intent in locating unwanted land uses, that it is has neglected to provide a full account of the ways in which accumulated environmental inequality then play into the social and the everyday—something that, in contrast, activist groups have centred on (Sze 2006). This insight moves us towards a recognition that environmental injustice arises not simply from unevenness in the spatial distribution of risk, from a politics of Cartesian geographical patterning, but from how this interacts with unevenness in socio-spatial distribution of vulnerability and wellbeing.

Spaces of Vulnerability and Wellbeing

The need to capture the interplay between vulnerability and the distribution of environmental bads has become more evident as the objects of attention within environmental justice discourse have diversified. For example, flooding is a threat to wellbeing that became part of environmental justice activism in the USA only after Hurricane Katrina devastated New Orleans in 2005 (Sze 2006; Pastor et al 2006). As a form of environmental risk for which there has been a deep engagement with concepts of vulnerability (Blaikie et al 1994; Cutter 1996; Pelling 2003) flooding clearly demonstrates the need to go beyond the socio-spatial patterning of risk in order to understand inequality. In the UK statistical environmental inequality analysis has been carried out to establish whether or not people who are experiencing multiple deprivation are more likely to live within the geographical boundaries of flood risk zones (Fielding and Burningham 2005; Walker et al 2003, 2007), showing that for coastal flooding in particular there is a strong spatial bias towards deprived people living in flood risk zones (Walker et al 2007). While this evidence of geographical patterning is partially revealing of inequality, its significance has to be seen in interaction with socio-spatial patterns in who is most vulnerable to flood impacts and how this vulnerability is being produced and reproduced for different people and communities. Here a catalogue of contributory dimensions of vulnerability need to be brought together—access to insurance, availability of resources to see through recovery, pre-existing health problems, infirmity, social isolation, the performance of emergency response and so on (Tapsell et al 2002; Walker et al 2007). In this light, inequality is not only a matter of the spatial distribution of risk—who lives on the floodplain and how they get to live there—but also of how each of these contributory dimensions to vulnerability also play out across space and time.

Similarly if we consider the environmental justice dimensions of greenspace, a form of environmental good, it is clear that the geographies involved again extend beyond simple proximity and socio-spatial patterning. A number of recent studies have used an environmental justice framing to consider "greenspace" (of different forms) as an environmental contribution to wellbeing. These studies have mapped the distribution of greenspaces in urban areas, analysing their prevalence in some parts of the city, near to some social groups, and their absence elsewhere (eg Fairburn, Walker and Smith 2005; Wolch and Wilson 2005). But evidently it is not simply presence that matters and that may or may not contribute to diminished wellbeing and the creation of a distributional injustice. The socio-cultural and scaled geography of meaning and significance also has to be part of the normative evaluations that are made (Heynen 2003). Greenspace is not an entirely uncontested and unproblematic "good thing" that contributes equally to wellbeing for all; rather there can be important cultural, gender and other differences in how particular forms of greenspace are viewed and the functions and services that these perform (Low, Taplin and Scheld 2006). It is also well recognised that there are many factors that can act as barriers to the access and use of greenspace for people in different social groups and contexts, such as fear of crime and of others, physical barriers to mobility, and conflicts between different uses and users (Gobster 1999; Risbeth and Finney 2006; Schmelzkopf 1995). It follows that how the meaning and impact of greenspaces on wellbeing shift across the city and over time (Brownlow 2006) may be *just* as significant as the geography of greenspace availability. In this light proximity is only one dimension of spatialised narratives of difference and inequality.

Space and Distribution Intertwined

Moving across the three examples used in this section of the chapter—pollution, flooding and greenspace—we can see that it is not just the socio-spatial patterning of the environment that matters to the distribution of outcomes and impacts on health and wellbeing. Other forms and scales of spatial relations are in interaction with this patterning, contributing to how vulnerability is constructed and wellbeing is diminished or improved. This has two implications for how the spatial is intertwined with environmental justice claims. First, it is clear that injustice, in terms of distributional outcomes, cannot be reduced simply and solely to tests of unequal spatial patterning and disproportionate proximity. Other distributional inequalities may compound these or, crucially, may constitute the basis for environmental injustice claims *even where* seemingly equal and even socio-spatial patterns of environmental goods or bads are observed (eg an "equal" distribution of pollution, flood risk or greenspace). Indeed overreliance on simple and uni-dimensional geography in environmental justice analysis may serve to obscure inequalities that are constituted and spatialised in different ways.

Second, as we move from concern to concern *and* from context to context, we can expect shifts in both the spatial relations that are seen to be significant and in the nature of justice claims being made. This is not simply because of the different material circumstances and situations involved, although these are important—Harvey (1996:6) argues that "different socio-ecological circumstances imply quite different approaches to the question of what is or is not just". It is also because there may be quite different understandings of the environmental goods and bads themselves—echoing Walzer's (1983:6) observations about the necessary pluralism of justice concepts—and because acts of claim-making are strategic and situated (Harvey 1996; Wenz 1988). For these combined reasons and as returned to in the conclusion, different constructions of the spatiality of distributional inequality will become more or less relevant and productive for actors in justice debates.

Geographies of Responsibility

While inequalities in the distribution of outcomes and impacts on wellbeing have to be central to a concern for justice in general, and for environmental justice in particular (Low and Gleeson 1998; Schlosberg 2007), questions of distributional justice may also centre on responsibility for the production of these outcomes. Distinctions can be made in justice theory (and everyday justice practice) between situations in which distributional inequalities are the consequences of the actions or informed choices made by the same people who are affected by them, and those where there is a dislocation between those benefiting from and suffering from patterns of distribution (Barry 1989; Wenz 1995). In particular, when harm or diminished wellbeing is experienced by already marginalised groups as a direct consequence of the actions of those that are more advantaged, then claims of injustice become particularly powerful. These questions of the relations between patterns of responsibility and patterns of outcome can have distinct spatialities to them that are a significant part of the normative evaluations that are made.

For example, in the case of waste the geography of responsibility and the spatial relations between sites of waste production, transformation and disposal have been important to catalysing environmental justice activism. The first generally recognised case of environmental justice protest in Warren County, North Carolina was stirred not *only* because the host community was predominantly black and poor (Shrader-Frechette 2002), but because the toxic soil waste to be disposed of was coming from 14 other counties where polychlorinated biphenyls (PCBs) had been illegally sprayed onto roadside soil (Bryant 2003).

This was not the waste of Warren County, but waste produced by others. Similarly in the UK, the first local protest action most explicitly using the language of environment justice, at Greengairs in Scotland in 1998 (Dunion 2003), was catalysed by a plan to dispose of the toxic waste of "others" in an extension to a major cluster of landfill sites. In this case the waste was not only coming from England, crossing a border that, at the time, was becoming increasingly suffused with political and cultural significance, but was being transported from Hertfordshire, one of the wealthier "home counties" near to London. Here the disconnected geographies of responsibility and outcome deeply mattered to the claim of injustice. The waste involved was not anonymous, but carried its identity with it as it travelled and crossed significant political boundaries. In this case a politicised inequality of flow, movement and responsibility intersected with an inequality of population proximity to landfill sites in the construction of an integrated justice claim.

Geographies of responsibility have emerged in a different form in establishing claims of injustice related to the socio-spatial distribution of air quality. Various studies in the UK have identified that the worst quality is typically found in the most deprived communities, both in terms of average concentration levels and exceedences of air quality standards (eg Mitchell and Dorling 2003; Walker et al 2003). While such distributional evidence may in its own right substantiate a claim of injustice, in particular when geographically coincident with heightened levels of vulnerability (as argued above), the spatiality of responsibility for poor air quality is also embroiled. Stevenson et al (1998) take the example of London, and argue that "clear" injustice arises because the poor air quality experienced in the most deprived areas of inner city London is the responsibility not of the people living in those areas—who have low levels of car ownership—but of those commuting in and out of the city to more wealthy suburbs and outlying towns. Mitchell and Dorling (2003) provide a similar analysis for Great Britain as a whole and conclude that while the poor, in general, do contribute to the worsening of air quality, wards with the very worst air quality were the poorest in the country and contributed the least to emissions—a situation which they conclude is "patently unjust". Here, as with waste flows, the spatialities of consumption and production both matter. The particulates and nitrogen oxides emitted from passing cars, accumulating in the atmosphere and inhaled into vulnerable lungs, are given a social and spatial identity that is disconnected from the communities experiencing unequal health outcomes.

There are many other examples of how the co- or dislocation of the consumption and production of environmental inequalities are central to justice claims, including international and global issues such as the transfer and disposal of hazardous waste and mitigation and adaptation to climate change (Ikeme 2003; Paavola and Adger 2002), both of which have been positioned within an environmental justice frame. How the spatialities of responsibility are conceived at such scales can be significant to the construction of competing justice claims and to the principles that are advocated for political and regulatory responses (Newell 2005). In campaigns and policies on climate change, greenhouse gas emissions are assigned a nation-state identity, through the construction of national emission inventories, the estimation of national totals and per capita indicators, and the assignment of national emission reduction targets. How nation-states are then blocked into regional or other groups and the extent to which the historical and the geographical are combined to take account of "legacy emissions" are central to the intense debates that have played out in international negotiations (Roberts and Parks 2007). For the regulation of international trade in hazardous and e-waste (Adeola 2000; Smith et al 2006), a nation-state responsibility is also assigned, with the Basel Convention requiring disposal of waste to take place within the national borders of where it is produced. A national identity is given to the waste as part of establishing what constitutes a just and equitable solution to dealing with it—even though when seen through a different scalar lens, that solution will ultimately involve a distribution of risk that is local and particular to a place, rather than national and collective in scope.

These examples leave us with a provisional view at least of how the simultaneous, interconnected analysis of the socio-spatialities of responsibility and impact can be a crucial part of environmental injustice claims. The geographies involved in mapping the distribution of responsibility and construction of injustice claims again shift from case to case; in some cases they are concerned with the spatiality of flows and the carrying of identity across politically or culturally significant boundaries; in others they are concerned with spatial fractures between the sites of consumption and production of environmental bads and goods using established hierarchic, but also potentially far more fluid notions of scaled comparison and difference (Newell 2005). Here again there are different constructions of space involved and opportunities for activists and institutions to work with the spatial in different and strategic ways.

Geographies of Recognition and Participation
The discussion so far has been concerned with different forms and parameters of distribution—of impacts, vulnerabilities, wellbeing, responsibilities—and how these are spatially constituted and interrelated. Developments in justice theory, however, have shown that to only be concerned with justice as distribution, to be locked into a

Rawlsian framework of need, desert and entitlement, is insufficient—both theoretically and for capturing the nature of justice as practised and argued over in everyday public life. Key here has been the work of Young (1990) and Fraser (1997) who, while following different lines of argument, have both sought to extend conceptions of justice in ways that focus attention on the processes through which distributional injustices are created and sustained (in this way both seek to supplement rather than replace distributional perspectives). Schlosberg (2004) draws on both theorists to argue that environmental justice in theory and praxis is "trivalent", integrating questions of distribution with those of participation and recognition in order to derive a more complete and satisfactory account. Accordingly he argues that:

> These notions and experiences of injustice are not competing notions, nor are they contradictory or antithetical. Inequitable distribution, a lack of recognition and limited participation all work to produce injustice and claims for injustice. (Scholsberg 2004:529)

He also persuasively shows in recent work (Schlosberg 2007) how all three concepts of justice are integrated in the arguments, discourses and principles of environmental justice activists in the USA and in global justice movements, and that they in this way accept "both the ambiguity and the plurality that come with such a heterogeneous discourse" (2007:5). Indeed he argues that "within the environmental justice movement, one simply cannot talk of one aspect of justice without it leading to another" (2007:73). Taking recognition and participation into our understanding of the nature of environmental justice in this way raises new questions about its intertwining with geography. In what way is space embroiled and interwoven with environmental justice as recognition and participation, as we have seen it is with justice as distribution?

Spaces of Misrecognition

Taking recognition first, there are a number of ways in which recognition, in the context of environmental justice, might be spatially constituted. At the core of misrecognition are cultural and institutional processes of disrespect, denigration, insult and stigmatisation, which devalue some people in comparison to others (Fraser 1997). While such devaluing of, for example gender, ethnic or racial groups, need not have an explicitly spatial expression, it is well recognised that there are circumstances in which the misrecognition of people can be entwined with and realised through the misrecognition of places. In the literature on socio-cultural understandings of environmental risk, notions of stigmatisation (drawing on Goffman 1963) have been used to explain why particular cases, usually of proposed development-producing technological risks, have generated particularly acute public resistance (Flynn, Slovic and Kunreuther 2001; Satterfield and Gregory 2002). Place stigmatisation, it is argued, can result from the siting of stigmatised technologies, such that positive senses of place are threatened and replaced with associations of danger, threat and degradation (Slovic, Flynn and Gregory 1994; Simmons and Walker 2005). It follows, as Pulido (1996) argues, that environmental justice mobilisations have often been seeking to reclaim denigrated places and place identities. Similarly, for Sze (2006:18) "environmental justice activism is about racial, geographic and local identity, as much at the same time as it is about a specific facility, issue or campaign".

Place stigmatisation and misrecognition are not however just the *product* of siting decisions, but also underlie the processes through which certain spaces get to be chosen for development in the first place. Once places, as well as people and communities, become "associated with trash" (Pellow 2002) they can then become the strategic or "natural destination" for further unwanted land uses. Accusations of environmental racism at the core of environmental justice in the US suggest deliberate strategic intent based on misrecognition of both people and places. Processes of land use planning that concentrate industrial activities, waste handling and energy generation together in "marked" places (literally so in terms of land use zonings), and that protect the environmental quality and land values of conservation and heritage areas, provide a less knowing and more institutionalised account of how recognition plays into the socio-spatial patterning of urban-industrial geography. Pulido (2000) provides an important move in this respect in analysing how "white privilege", a highly structural and spatial form of racism, has both shaped the urban landscape and created distinct but functionally related clean residential suburbs and polluted industrial zones (see Leichenko and Solecki 2008 for a related analysis of gated communities). Similar institutionalised understandings of misrecognition can be used to explain, in part at least, why the immediate "doorstep" environment of marginalised places—the streets and neighbourhoods of daily life for the poor or particular ethnic groups—becomes neglected and poorly served by the mundane environmental services of street cleaning and maintenance (Lucas et al 2004; Hastings et al 2005); as captured in Scotland by the term "environmental incivilities" (Curtice et al 2005). Marked people in marked places become expected to live with incivilities and blamed for not looking after their own environment, with such institutionalised assumptions shaping where effort by the state to address problems is and is not deployed.

Place stigmatisation in which people and places are associated is not the only way in which recognition is spatialised. People moving into and through spaces and environments with which they are disassociated and culturally disconnected can also bring misrecognition into claims of environmental injustice. In the UK one of the first connections made between issues of race, ethnicity and the environment related to who was visible in and making use of rural spaces (Agyeman 1990). The lack of "black faces in the countryside" became a particular focus of the Black Environment Network in the 1980s, and has more recently been given attention by the Commission for Racial Equality as well as organisations responsible for countryside management. While a number of historical and contemporary processes can be seen to be at work in reproducing a predominantly white British rurality (Cloke and Little 1997), for minority communities the ways in which culturally embedded misrecognition became more acute as they moved from urban into rural spaces was central to how they felt excluded from rural environments. Drawing from this example, there are potentially many other ways in which understandings of the geography of identity can bring insights into how the spatiality of cultural and institutional misrecognition underpins the maldistribution of environmental goods and bads.

Spaces of Fair Process

Turning finally to justice as procedure, there is a sense in which a call or demand for more democracy, openness and inclusion in processes of decision-making is about enabling access to spaces, and flows between spaces, that have previously been restricted (Barnett and Low 2004). In this way a lack of procedural justice is intimately wrapped up with a closed geography of information, access and power—and procedural fairness with a fluidity of movement of people, ideas and perspectives across the boundaries of institutions and between differentiated elite and lay spaces, creating open rather than constrained networks of interaction and deliberation. The degree to which such fluidity and interaction is *genuinely* achieved and has influence is through the crucial test of procedural fairness—as realised rather than discursively represented. The real-world geography of flows, encounters and power relations is an important part of that test. Examples of how the spatial factors into the realities of "just" procedure and process include the following: the ways in which access to the "open provision" of web-based environmental information and the deliberative possibilities of virtual participation (Zavestoski, Shulman and Schlosberg 2006) are in practice spatially and socially differentiated; the ways that access to resources and the time–space constraints of everyday life limit abilities to be present in participatory spaces, from local meetings to international negotiations (Barnes et al 2003; Roberts and Parks 2007); and how strategic behaviour operates within and outside of the formal spaces of decision-making processes (Bickerstaff and Walker 2005; Sherlock, Kirk and Reeves 2004).

The marking out of democratic space is also part of, but problematic within, prescriptions for how procedural environmental justice should be achieved (Lake 1996). Many calls for procedural justice assert that those who are most affected by decisions should have particular rights to be involved and have their voices heard on a fully informed basis (Hampton 1999). However, this begs the question of how "those who are most affected" should be defined. Spatial boundaries, delineated on political, environmental or cultural grounds, are often involved in such a definition but are rarely unproblematic. Hunold and Young (1998), writing within an environmental justice framing, provide the most thorough attempt to define what fairness should constitute in decision-making related to the siting of hazardous facilities, but fail to grapple sufficiently with the complexities of geography that can be involved. For example, a key part of their prescription is that "siting policy should be made on the basis of a fairly large unit of review—at the state or regional level—or the decision about where a site is located will already have been made" (1998:91), a provision that fails to recognise the problems involved in, for example, selecting sites that sit near to and generate impacts that transgress the political logic of state or regional boundaries. Somewhat ironically, such problems of spatial definition become all the more acute the more that power to determine or negotiate decision outcomes is passed to "the community" and/or mechanisms of resourcing involvement or compensating impacts are deployed (as is becoming increasingly advocated in siting policy; Lesbirel and Shaw 2005). For example, in the UK the principle of "fairness with respect to procedures, communities and future generations" (Committee on Radioactive Waste Management 2007:13) has been stated as central to the process to be used to decide where to site a deep geological repository for the disposal of nuclear waste. An innovative package of volunteering, resourcing of community involvement in negotiations, and compensation for the eventual selected host has been proposed, but questions of spatial definition are deeply problematic (as the Committee itself recognises). How should "volunteer communities" be defined and enabled to enter into the process? How can a focus on empowering host communities be reconciled with the risks that would be experienced by communities along transport corridors? Over what area should compensation be negotiated when risks and associated stigma impacts arguably extend far beyond the immediate locality? Such spatially orientated dilemmas are common to other situations in which justice is an explicit part of environmental decision processes— for example, the negotiation of rights to indigenous genetic materials under the Convention on Biological Diversity (Vermeylen 2007)—and demonstrates the very real ways in which the construction of space is wrapped up in the determination of fair process and who is included and excluded from the environmental justice that is performed.

Conclusion

The remit of this chapter has been intentionally wide ranging in that I have endeavoured to identify the multiple ways in which geography, and specifically the spatial, is intertwined with a pluralised understanding of the scope and meaning of environmental justice. In the course of working through various distributional dimensions of impacts and responsibilities, through justice conceived as recognition and fair procedure and through a diversity of examples of socio-environmental concerns, this explorative process has encountered multiple spatialities of different things, of different forms, constructed at different scales. Environmental justice has been intertwined with the spatiality of people of different ethnicities, ages and genders; industrial installations, traffic and greenspaces; waste categories and molecules of pollution; perceptions, identities and meanings. Space has taken different forms— Cartesian space; political and democratic space; institutional space; spaces of identity, place and community; dynamic spaces of flows; and movement between spaces and across boundaries. Space has been organised and constructed at different scales: local proximate space, the body, the community, the region, and the nation-state. This substantial but inevitably partial listing is sufficient to demonstrate that the uni-dimensional and simple distributional geography of proximity that has characterised the enrolment of space within "first-generation" understandings of environmental justice forms only one part of a far more topologically involved landscape of socio-environmental relations. Such multidimensionality is not intrinsically a good thing (more is not necessarily better[2]), and clearly not all dimensions are necessarily equal in their significance or prevalence. But being open and receptive to diversity and plurality, rather than assuming that certain conventions of justice and spatiality will always be present or dominant, is, I would argue, necessary to do justice to (and in) a rapidly evolving field.

A number of conclusions and implications flow from this analysis. First, it lends support to Harvey's (1996) argument that justice and geography matter together; that they interrelate and are co-constructed as claims of inequality and injustice are put forward. It follows that how environmental justice is conceived will bring forward certain understandings of space and hide others; and that how space is conceived will open up certain avenues for claiming environmental injustice, and close down others. As Kurtz (2002) and Towers (2000) argue, specifically in relation to scalar framings, this pluralism and fluidity mean that the politics of scale—or I would argue a broader politics of space—is significant in the way that environmental justice disputes are played out. Different forms and scales of space are in this sense a strategic resource and just as "different groups will resort to different conceptions of justice to bolster their position" (Harvey 1996:398), so will different groups work with different understandings of the spatiality of the issues at hand.

Second, if, as Schlosberg (2007) argues, different understandings of justice—as distribution, recognition and participation—are simultaneously applied and integrated within the discourses of environmental justice activists, so we might expect to observe multiple spatialities at work. This is not only a matter of the scalar shifting and interlinking that various analyses have identified in the tactics of activist groups (Davies 2006; Kurtz 2002), but a wider set of possibilities for simultaneously working with different spatial conceptions of impacts, vulnerabilities, responsibilities, recognition and participation, and for integrating these together. Leitner, Sheppard and Sziarto (2008) have recently made a similar argument about the multiple spatialities of contentious politics more generally, suggesting that there is a productive opportunity to tie the analysis of the spatialities of environmental justice activism to wider debates within the discipline. In particular, we might explore how different spatialities are being tied in congruent and supportive ways to produce more rather than less resilient multidimensional environmental justice discourses.

Third, a limitation of the analysis in this chapter is that is has been unable to represent the great diversity of political and cultural contexts into which an environmental justice frame has travelled across the world (Schroeder et al 2008; Walker and Bulkeley 2006). However, it is possible to speculate what implications the arguments developed here might have for our understanding of how environmental justice translates. If the spatiality of environmental justice was simply distributional and proximate, then this formulation and the practices and discourses that flow from it could travel relatively untouched from context to context. A circle mapped around an industrial plant and a population statistically analysed in Los Angeles, Lancaster, Johannesburg or Mexico City is ontologically stable, even if the details of data and socio-environmental categories may change. However, if the spaces that matter are not Cartesian in form but those of place identity, community, process and procedure, or if the meanings and values given to social and environmental spaces are socio-culturally rather than statistically defined, then we should expect both the meaning and spatiality of environmental justice to shift and reform as the framing travels and translates. For debates about the (im)possibilities of universalism in environmental justice theory and praxis (Harvey 1996;

Schlosberg 2007; Walker and Bulkeley 2006; Williams and Mawdsley 2006) this necessarily supports a pluralistic perspective, even if common core issues and processes can be observed across different parts of the world (Schroeder et al 2008).

Finally, the foregoing analysis inherently makes the case for a new, methodologically diverse and theoretically pluralised stream of geographical scholarship on environmental justice. This chapter has drawn from a diversity of human and environmental geography scholarship that, while not necessarily positioned within an environmental justice framing, has provided insights into the nature of socio-environmental relations, as well as into how justice and space are intertwined. In future research there is scope for a more thorough analysis of the spatiality of environmental justice within different socio-environmental and political contexts, for exploration of the implications of multiplicity and diversity that have been suggested in this conclusion, and for a closer engagement by geographers with recent developments in environmental justice theory.

Acknowledgements

I am grateful for the insightful comments and suggestions of two referees. Collaborative work over the past 8 years with Karen Bickerstaff, Malcolm Eames, Jon Fairburn, Harriet Bulkeley and Gordon Mitchell has contributed greatly to my thinking about the geography of environmental justice.

Endnotes

[1] He also introduces the capability framework in his most recent work (Scholsberg 2007) in part as it can be used to contain multiple notions of justice. This is not included in the analysis of the chapter but does provide intriguing possibilities for further development in the context of environmental justice (Holland 2008; Walker 2009).
[2] I am grateful to a referee for this challenging insight.

References

Adeola F O (2000) Cross-national environmental injustice and human rights issues. *American Behavioural Scientist* 43(4):686–706
Agyeman J (1990) Black people in a white landscape: Social and environmental justice. *Built Environment* 16(3):232–236
Agyeman J and Evans B (2003) Just sustainability: The emerging discourse of environmental justice in Britain. *The Geographical Journal* 170(2):155–164
Barnes M, Newman J, Knops A and Sullivan H (2003) Constituting "the public" in public participation. *Public Administration* 81(2):379–399
Barnett C and Low M (eds) (2004) *Spaces of Democracy: Geographical Perspectives on Citizenship, Participation and Representation.* London: Sage
Barry B (1989) *Theories of Justice.* Berkeley: University of California Press
Benford R (2005) The half-life of the environmental justice frame: Innovation, diffusion and stagnation. In D N Pellow and R J Brulle (eds) *Power, Justice and the Environment: A Critical Appraisal of the Environmental Justice Movement* (pp 37–54). Cambridge, MA: MIT Press

Bickerstaff K and Walker G P (2003) The place(s) of matter: Matter out of place—public understandings of air pollution. *Progress in Human Geography* 27(1):45–67

Bickerstaff K and Walker G P (2005) Shared visions, unholy alliances: Power, governance and deliberative processes in local transport planning. *Urban Studies* 42(5): 2123–2144

Blaikie P, Cannon T, Davis I and Wisner B (1994) *At Risk: Natural Hazards, Peoples Vulnerability and Disasters*. London: Routledge

Bowen W (2002) An analytical review of environmental justice research: What do we really know? *Environmental Management*. 29(1):3–15

Bowen W M and Wells M V (2002) The politics and reality of environmental justice research: A history and considerations for public administrators and policy makers. *Public Administration Review* 62(6):688–698

Brown P (1995) Race, class and environmental health: A review and systemisation of the literature. *Environmental Research* 69:15–30

Brownlow A (2006) An archaeology of fear and environmental change in Philadelphia. *Geoforum* 37:227–245

Bryant B (2003) History and issues of the Environmental Justice Movement. In Visgilio G R and Whitelaw D M (eds) *Our Backyard, A Quest for Environmental Justice* (pp 3–24) Lanham: Rowman and Littlefield

Buckingham-Hatfield S, Reeves D and Batchelor A (2005) Wasting women: The environmental justice of including women in municipal waste management. *Local Environment* 10(4):427–444

Bullard R D (1999) Dismantling environmental justice in the USA. *Local Environment* 4(1):5–20

Bullard B, Mohai P, Saha R and Wright B (2007) *Toxic Wastes and Race at Twenty: 1987–2007*. Cleveland: United Church of Christ Justice and Witness Ministries

Buzzelli M (2007) Bourdieu does environmental justice? Probing the linkages between population health and air pollution epidemiology. *Health & Place* 13(1):3–13

Carey H H (2002) Forest management and environmental justice in Northern New Mexico. In K Mutz, C B Gary and S K Douglas (eds) *Justice and Natural Resources: Concepts, Strategies, and Applications* (pp 209–224). Washington DC: Island Press

Chaix B, Gustafsson S, Jerret M, Kristerson H, Lithman T, Boalt A and Merlo J (2006) Childrens exposure to nitrogen dioxide in Sweden: Investigating environmental justice in an egalitarian country. *Journal of Epidemiology and Community Health* 60:234–241

Charles A and Thomas H (2007) Deafness and disability—forgotten components of environmental justice: Illustrated by the case of Local Agenda 21 in South Wales. *Local Environment* 12(3):209–221

Cloke P and Little J (eds) (1997) *Contested Countryside Cultures*. London: Routledge

Committee on Radioactive Waste Management (2007) Implementing a partnership approach to radioactive waste management. Report to Governments, April 2007, CoRWM document 2146. http://www.corwm.org.uk/ Accessed 2 April 2008

Curtice J, Ellaway A, Robertson C, Morris G, Allardice G and Robertson R (2005) *Public Attitudes and Environmental Justice in Scotland*. Edinburgh: Scottish Executive

Cutter S (1995) The forgotten casualties: Women, children and environmental change. *Global Environmental Change* 5(3):181–194

Cutter S (1996) Vulnerability to environmental hazards. *Progress in Human Geography*, 20(4):529–539

Davies A (2006) Environmental justice as subtext or omission: Examining discourses of anti-incineration campaigning in Ireland. *Geoforum* 37(5):708–724

Dobson A (1998) *Justice and the Environment*. Oxford: Oxford University Press

Dunion K (2003) *Trouble Makers. The Struggle for Environmental Justice in Scotland.* Edinburgh: Edinburgh University Press

Fairburn J, Walker G and Smith G (2005) *Investigating Environmental Justice in Scotland—Links Between Measures of Environmental Quality and Social Deprivation.* Report UE4(03)01. Edinburgh: Scottish and Northern Ireland Forum for Environmental Research

Fan M F (2006) Environmental justice and nuclear waste conflicts in Taiwan. *Environmental Politics* 15(3):417–434

Fielding J and Burningham K (2005) Environmental inequality and flood hazard. *Local Environment* 10(4):379–410

Flynn J, Slovic P and Kunreuther H (eds) (2001) *Risk, Media and Stigma: Understanding Public Challenges to Modern Science and Technology.* London: Earthscan

Fraser N (1997) *Justice Interruptus: Critical Reflections on the "Postsocialist" Condition.* New York: Routledge

Gee G and Payne-Sturges D C (2004) Environmental health disparities: A framework integrating psychosocial and environmental concepts. *Environmental Health Perspectives* 112(7):1645–1653

Getches D and Pellows D (2002) Beyond "traditional" environmental justice. In Mutz K, Gary C B and Douglas S K (eds) *Justice and Natural Resources: Concepts, Strategies, and Applications.* Washington DC: Island Press

Gobster P (1999) Urban parks as green walls or green magnets? Interracial relations in neighbourhood boundary parks. *Landscape and Urban Planning* 41:43–55

Goffman E (1963) *Stigma: Notes on the Management of Spoiled Identity.* Harmondsworth: Penguin

Hampton G (1999) Environmental equity and public participation. *Policy Sciences* 32(2):163–174

Harvey D (1996) *Justice, Nature and the Geography of Difference.* Oxford: Blackwell

Hastings A, Flint J, McKenzie C and Mills C (2005) *Clearing up Neighbourhoods.* Bristol: Policy Press

Heynen N (2003) The scalar production of injustice within the urban forest. *Antipode* 35(5):980–998

Hillman M (2006) Situated justice in environmental decision-making: Lessons from river management in Southeastern Australia. *Geoforum* 37(5):695–707

Holifield R (2001) Defining environmental justice and environmental racism. *Urban Geography* 22:78–90

Holifield R (2004) Neoliberalism and environmental justice in the US EPA: Translating policy into managerial practice in hazardous waste remediation. *Geoforum* 35:285–297

Holland B (2008) Justice and the environment in Nussbaum's "capabilities approach": Why sustainable ecological capacity is a meta-capability. *Political Research Quarterly* 61(2):319–332

Honneth A (2001) Recognition or redistribution: Changing perspectives on the moral order of society. *Theory, Culture and Society* 18(2–3):43–55

Hunold C and Young I M (1998) Justice, democracy and hazardous siting. *Political Studies* 46(1):82–95

Hurley A (1995) *Environmental Inequalities: Class, Race and Industrial Pollution in Gary, Indiana, 1945–1980.* Chapel Hill, NC: University of North Carolina Press

Ikeme J (2003) Equity, environmental justice and sustainability: Incomplete approaches in climate change politics. *Global Environmental Change* 13:195–206

Kuehn R R (1997) An analysis of the compatibility of quantitative risk assessment with the principles of environmental justice in the United States. *Risk Decision and Policy* 2(3):259–276

Kurtz H E (2002) The politics of environmental justice as the politics of scale: St James Parish, Louisiana and the Shintech siting controversy. In A Herod and M W Wright (eds) *Geographies of Power: Placing Scale* (pp 249–273). Oxford: Blackwell

Kurtz H E (2007) Gender and environmental justice in Louisiana: Blurring the boundaries of public and private spheres. *Gender Place and Culture* 14(4):409–426

Laird S, Cunningham A B and Lisinge E (2000) One in ten thousand? The Cameroon case of Ancistrocladus korupensis. In C Zerner (ed) *People, Plants and Justice: The Politics of Nature Conservation* (pp 345–373). New York: Columbia University Press

Lake R (1996) Volunteers, NIMBYs, and environmental justice: Dilemmas of democratic practice. *Antipode* 28(2):160–174

Leichenko R and Solecki W (2008) Consumption, inequity, and environmental justice: The making of new metropolitan landscapes in developing countries. *Society and Natural Resources* 21(7):611–624

Leitner H, Sheppard E and Sziarto K M (2008) The spatialities of contentious politics. *Transactions of the Institute of British Geographers* 33(2):157–172

Lesbirel S H and Shaw D (eds) (2005) *Managing Conflict in Facility Siting: An International Comparison.* Cheltenham: Edward Elgar

Liu F (2001) *Environmental Justice Analysis: Theories, Methods and Practice.* Boca Raton: Lewis Publishers

London L (2003) Human rights, environmental justice, and the health of farm workers in South Africa. *International Journal of Occupational and Environmental Health* 9(1):59–68

Longhurst R (2001) *Bodies: Exploring Fluid Boundaries.* London: Routledge

Low N and Gleeson B (1998) *Justice, Society and Nature: An Exploration of Political Ecology.* London: Routledge

Low S, Taplin D and Scheld S (2006) *Rethinking Urban Parks: Public Space and Cultural Diversity.* Austin: University of Texas

Lucas K, Walker G, Eames M, Fay H and Poustie M (2004) *Environment and Social Justice: Rapid Research and Evidence Review,* Sustainable Development Research Network, London: Policy Studies Institute

Mitchell G and Dorling D (2003) An environmental justice analysis of British air quality *Environment and Planning A* 35(5):909–929

Mutz K, Gary C B and Douglas S K (eds) (2002) *Justice and Natural Resources: Concepts, Strategies, and Applications.* Washington, DC: Island Press

Newell P (2005) Race, class and the global politics of environmental inequality. *Global Environmental Politics* 5(3):70–94

Omer I and Or U (2005) Distributive environmental justice in the city: Differential access in two mixed Israeli cities. *Tijdschrift Voor Economische En Sociale Geografie* 96(4):433–443

O'Neill M, Jerrett S, Kawachi I, Levy J I, Cohen A J, Gouveia N, Wilkinson P, Fletcher T, Cifuentes L and Schwartz J (2003) Health, wealth and air pollution: Advancing theory and methods, *Environmental Health Perspectives* 111(16):1861–1870

Paavola J and Adger W N (2002) Justice and adaptation to climate change. Tydnall Centre Working Paper No 23. Norwich: University of East Anglia

Pastor M, Bullard R D, Boyce J K, Fothergill A, Morello-Frosch R and Wright B (2006) *In the Wake of the Storm: Environment, Disaster and Race after Katrina.* New York: Russell Sage Foundation

Pearce J, Kingham S and Zawar-Reza P (2006) Every breath you take? Environmental justice and air pollution in Christchurch, New Zealand. *Environment and Planning A* 38:919–938

Pelling M (2003) *The Vulnerability of Cities: Social Resilience and Natural Disaster*. London: Earthscan

Pellow D (2002) *Garbage Wars: The Struggle for Environmental Justice in Chicago*. Cambridge, MA: MIT Press

Petts J (2005) Enhancing environmental equity through decision making: Learning from waste management. *Local Environment* 10(4):397–410

Pulido L (1996) A critical review of the methodology of environmental racism research. *Antipode* 28(2):142–159

Pulido L (2000) Rethinking environmental racism: White privilege and urban development in Southern California. *Annals of the Association of American Geographers* 90(1):12–40

Rishbeth C and Finney N (2006) Novelty and nostalgia in urban greenspace: Refugee perspectives. *Tijdschrift voor Economische en Sociale Geografie* 97(3):281–295

Roberts J T and Parks B C (2007) *A Climate of Injustice: Global Inequality, North–South Politics and Climate Policy*. Cambridge, MA: MIT Press

Satterfield T and Gregory R (2002) The experience of risk and stigma in community contexts. *Risk Analysis* 22:347–358

Schmelzkopf K (1995) Urban community gardens as contested space. *Geographical Review* 85(3):364–381

Scholsberg D (2002) *Environmental Justice and the New Pluralism*. Oxford: Oxford University Press

Scholsberg D (2004) Reconceiving environmental justice: Global movements and political theories. *Environmental Politics* 13(3):517–540

Scholsberg D (2007) *Defining Environmental Justice: Theories, Movements and Nature*. Oxford: Oxford University Press

Sherlock K L, Kirk E A and Reeves A D (2004) Just the usual suspects? Partnerships and environmental regulation. *Environment and Planning C: Government and Policy* 22(5):651–666

Shrader-Frechette K (2002) *Environmental Justice: Creating Equality and Reclaiming Democracy*. Oxford: Oxford University Press

Schroeder R (2000) Beyond distributive justice: Environmental justice and resource extraction. In C Zerner (ed) *People, Plants and Justice: The Politics of Nature Conservation* (pp 52–64). New York: Columbia University Press

Schroeder R, Martin K, Wilson B and Debarati S (2008) Third World environmental justice. *Society and Natural Resources* 21(7):547–555

Simmons P and Walker G P (2005) Technological risk and sense of place: Industrial encroachment on place values. In A Boholm and R Lofstedt (eds) *Facility Siting: Risk, Power and Identify in Land Use Planning* (pp 90–106). London: Earthscan

Slovic P, Flynn J and Gregory R (1994) Stigma happens—social-problems in the siting of nuclear waste facilities. *Risk Analysis* 14:773–777

Smith D (1994) *Geography and Social Justice*. Oxford: Blackwell

Smith T, Sonnenfeld D A and Pellow D N (eds) (2006) *Challenging the Chip: Labor Rights and Environmental Justice in the Global Electronics Industry*. Philadelphia, PA: Temple University Press

Stephens C, Bullock S and Scott A (2001) Environmental justice: Rights and means to a healthy environment for all. Special Briefing Paper 7. Economic and Social Research Council (ESRC) Global Environmental Change Programme. Brighton: ESRC Global Environmental Change Programme, University of Sussex

Stevenson S, Stephens C, Landon M, Pattendon S, Wilkinson P and Fletcher T (1998) Examining and inequality and inequity of car ownership and the effects of pollution and health outcomes such as respiratory disease. *Epidemiology* 9(4):S29

Sze J (2006) Toxic soup redux: Why environmental racism and environmental justice matter after Katrina. *Understanding Katrina: Perspectives from the Social Sciences.* http://understandingkatrina.ssrc.org Accessed 19 March 2009

Tapsell S M, Penning Rowsell E C, Tunstall S M and Wilson T L (2002) Vulnerability to flooding: Health and social dimensions. *Philosophical Transactions of the Royal Society A* 360:1511–1525

Targ N (2005) Running an empty: Transport, social exclusion and environmental justice. *Journal of the American Planning Association* 71(4):456–457

Towers G (2000) Applying the political geography of scale: Grassroots strategies and environmental justice. *Professional Geographer* 52(1):23–26

Vermeylen S (2007) Contextualizing "fair" and "equitable": The San's reflections on the Hoodia benefit-sharing agreement. *Local Environment* 12(4):423–436

Walker G (2009) Environmental justice and normative thinking. *Antipode* 41(1):203–205

Walker G P and Bulkeley H (2006) Geographies of environmental justice. *Geoforum* 37(5):655–569

Walker G, Burningham K, Fielding J, Smith G, Thrush D and Fay H (2007) *Addressing Environmental Inequalities: Flood Risk.* Science Report SC020061. Bristol: Environment Agency

Walker G, Mitchell G, Fairburn J and Smith G (2003) *Environmental Quality and Social Deprivation. Phase II: National Analysis of Flood Hazard, IPC Industries and Air Quality.* R&D Project Record E2-067/1/PR1. Bristol: The Environment Agency

Walzer M (1983) *Spheres of Justice.* Berkeley: University of California Press

Wenz P S (1988) *Environmental Justice.* Albany NY: SUNY Press

Wenz P S (1995) Just garbage. In L Westra and P S Wenz (eds) *Faces of Environmental Racism: Confronting Issues of Global Justice* (pp 57–71). Lanham: Rowman and Littlefield

Williams G and Mawdsley E (2006) Postcolonial environmental justice: Government and governance in India. *Geoforum* 37:660–670

Williams O (2005) Food and justice: The critical link to healthy communities. In Pellow D N and Brulle R J (eds) *Power, Justice and the Environment: A Critical Appraisal of the Environmental Justice Movement* (pp 117–130). Cambridge, MA: MIT Press

Wolch J and Wilson J P (2005) Parks and park funding in Los Angeles: An equity-mapping analysis. *Urban Geography* 26(1):4–35

Young I (1990) *Justice and the Politics of Difference.* Princeton, NJ: Princeton University Press

Zavestoski S, Shulman S and Schlosberg D (2006) Democracy and the environment on the Internet: Electronic participation in regulatory rulemaking. *Science, Technology and Human Values* 31(4):383–408

Zimmerman R (1994) Issues of classification in environmental equity: How we manage is how we measure. *Fordham Urban Law Journal* 21:633–669

Chapter 2

Actor-network Theory as a Critical Approach to Environmental Justice: A Case against Synthesis with Urban Political Ecology

Ryan Holifield

Introduction

Research on environmental justice occupies an ambivalent place in contemporary critical geography. On the one hand, the topic continues to generate considerable interest—new articles, books, and papers on environmental justice appear frequently,[1] and a glance at job listings suggests that environmental justice specialists remain in high demand. Once confined largely to quantitative studies of associations between selected environmental hazards—especially sources of toxic pollution—and demographic data on income and race, environmental justice research has branched out in the past two decades to tackle a much wider range of issues (eg, Sze and London 2008). It has also moved beyond its former focus on the United States, now addressing environmental inequities in countries throughout the world (Schroeder et al 2008; Stephens 2007; Walker and Bulkeley 2006). In numerous ways, geographic research on environmental justice and inequality continues to thrive.

On the other hand, some critical geographers and scholars in related fields have hinted that environmental justice research has stagnated. Some contend that it has remained empiricist in orientation, engaging to an exceptionally limited degree with theory—critical or otherwise (eg, Schweitzer and Stephenson 2007). Although environmental justice research has never been as disengaged from social, political, and geographic theory as some have suggested (eg, Buzzelli and Veenstra 2007; Kurtz 2003; Pulido 2000; Schlosberg 1999; Taylor 2000), empiricist studies of distributions of hazards or environmental justice activism—whether quantitative or qualitative—have nonetheless dominated the field. Studies that document and quantify environmental inequalities at different sites and scales provide valuable information, but they may have also contributed to the perception that environmental justice research is of marginal intellectual interest.

In recognition of both the continuing relevance of environmental inequality and the need for more theoretically informed environmental justice research, the relatively young field of urban political ecology (UPE)—the Marxist variety in particular—has emerged as an important and influential theoretical framework for environmental justice studies. Through the work of Swyngedouw and Heynen (2003) and others, Marxist UPE has not only laid out a clear theoretical and political agenda for addressing environmental injustice, but has also generated a rich and growing body of empirical research (see also Braun 2005a; Keil 2003).[2]

Given the apparent promise of Marxist UPE, why propose actor-network theory (ANT)—and specifically the version articulated by Bruno Latour—as an alternative critical approach to environmental justice studies? ANT, an orientation to social scientific research that developed within the field of science studies, has attracted considerable attention for its interest in the agency of nonhumans. ANT is already well established as an important influence on Marxist UPE, which prominently cites ANT scholars like Latour, Michel Callon, and John Law (eg, Swyngedouw 2004). Despite this growing influence, however, Latour's version of ANT has developed a reputation among critical scholars as a status quo approach that ignores inequalities, differences, and power relations; focuses its attention not on marginalized communities but on scientists and bureaucrats; produces only descriptions rather than powerful theoretical explanations; and remains stubbornly allergic to critique (eg, Haraway 1997; Hartwick 2000; Rudy 2005; Swyngedouw and Heynen 2003).[3] If critical geography has already taken the baby (attention to nonhumans) from ANT and thrown out the bathwater, then why make the case for ANT as an *alternative* approach to environmental justice research—and a critical one at that?

In the wake of Latour's (2004, 2005) recent books, which we might regard as either clarifications or revisions of ANT, critical urban political ecologists interested in conducting research on environmental justice would do well to take another look at ANT. I argue, first, that Marxist UPE's ambivalent orientation towards ANT—embracing hybrid networks and nonhuman agents, but rejecting other aspects of the approach as uncritical and insufficiently explanatory—rests on a particular reading of Latour, to which I offer an alternative. Second, instead of attempting to develop an urban political ecology that synthesizes Marxism and ANT (Castree 2002; Gareau 2005; Perkins 2007), I contend that critical scholars should keep the two separate. My argument is not that Latour's approach[4] invalidates or replaces Marxist

UPE, but that an actor-network approach to environmental justice research would enable us to do something different: to pursue questions that Marxist approaches might not open up. Specifically, instead of explaining inequalities by contextualizing and situating them, actor-network approaches turn our attention to the forms and standards that make it possible to circulate new associations of entities, to generalize social orders, and to situate actors within a social context—that is, to socialize them in particular ways. Finally, contrary to claims that ANT is "politically useless" and irredeemably inferior to Marxist approaches (Rudy 2005:119), I also maintain that an actor-network approach has the potential both to be politically relevant to environmental justice struggles and to deepen the democratic agenda of UPE.

An Environmental Justice Case Study

To illustrate contrasts between Marxist and actor-network approaches to environmental justice and injustice, I will briefly introduce a case explored in greater detail elsewhere: the assessment of human health risk at the St Regis Paper Company Superfund hazardous waste site in Cass Lake, Minnesota (Holifield 2007). I will present the case study here and, later in the chapter, revisit it twice: once with suggestions of what a Marxist UPE approach could bring to its analysis, and once with a brief exploration of what an actor-network approach could add. Although the site is not "urban" in the conventional sense—the 2000 US Census showed a population of 860 for Cass Lake—the dynamics of the case nonetheless point to new avenues of inquiry at more recognizably urban sites. The St Regis site, which hosted a wood treatment facility from the late 1950s to the early 1980s, was named in 1984 to the US Environmental Protection Agency's (EPA) National Priorities List; as a result, its contaminated soil and groundwater became eligible for long-term remediation under the federal Superfund program. In the second half of the 1980s, Champion International Corporation bought and closed the facility, set up a groundwater treatment system with extraction and monitoring wells, and moved contaminated soil to a large "vault" across the street from the former facility grounds.

The site became controversial in the early 1990s. First, data from the monitoring wells indicated that a plume of contaminated groundwater was migrating past the system of extraction wells toward surface waters and private wells to the east. Second, scientists working with local agencies discovered that the site's residual soil—the soil left behind after the top layer had been excavated and moved to the vault—had only been inspected visually, and had never been sampled and analyzed in a laboratory. Third, although the site is located completely within the territorial boundaries of an American Indian reservation—

the Leech Lake Reservation—EPA had delegated oversight to the state of Minnesota. The Leech Lake Band of Ojibwe contended not only that EPA had violated its own Indian Policy by delegating a site on tribal land to a state agency, but also that Minnesota's oversight of the remediation had been exceptionally lax, in effect handing the power to monitor the site's safety to the liable owners (Minnesota Sea Grant and Natural Resources Research Institute 2002). At this point residents and community leaders began to connect local health problems with site contamination and to identify this evident neglect—both of the site and of the Band's tribal sovereignty—as a case of environmental injustice.

Starting in the mid-1990s, the Leech Lake Reservation's Division of Resource Management began seeking to have human health and ecological risk assessments conducted for the site. By this time EPA had established quantitative risk assessments as the preferred means to determine whether remedial action—or, in this case, *new* remedial action—is warranted at a site. Consequently, risk assessments would be a precondition for EPA to order any new long-term cleanup action at the site. Under Minnesota's oversight, such risk assessments had never taken place.

However, at least two new developments complicated the prospects for risk assessment at St Regis. On the one hand, in response to pressure from regulated industries, EPA (1996) introduced a new policy under the Clinton administration that both allowed and encouraged liable site owners with the financial means to conduct risk assessments themselves (Powell 1999). The new policy formally overturned an older policy that had required the agency to conduct all Superfund risk assessments itself. Although EPA would continue to oversee risk assessments, the new policy placed considerable power over monitoring, sampling, and analysis into the hands of liable firms, including the owners of the St Regis site.

On the other hand, supported in part by new federal funds and policies for tribal governance and environmental protection, risk assessors had begun to develop customized assessments that would address risks to tribal traditional lifeways within American Indian reservations (Harris and Harper 1997). The latter, which challenged the universal applicability of EPA's standard default assumptions about human exposure, were designed to reflect both the disproportionate risks that tribal members could face as a consequence of practicing traditional ways of life—subsistence fishing, hunting, and gathering; sweat lodges; and so on—and the treaty rights of members to engage safely in these practices within reservation boundaries. At St Regis, tribal leaders and personnel regarded risk assessments that would take tribal distinctiveness into account as one important means that could help rectify the injustices associated with the site. However, EPA's new policy handed considerable power to design and carry out the risk assessments to the liable company.

Marxist Urban Political Ecology and Environmental Justice

Marxist UPE would provide one quite plausible conceptual framework for explaining the environmental injustices in the St Regis case. Marxist UPE aims to contextualize the production of uneven socio-environmental landscapes within the broader dynamics of capitalist urbanization (Braun 2005a; Keil 2003). In doing so, it brings together multiple theoretical and topical threads, including political ecology—especially "First World political ecology" (McCarthy 2002; Schroeder 2005)—critical urban theory, theories of the social production of nature, and the political economy of environmental justice discourse. Marxist UPE explains contemporary environmental inequalities as products or at least reflections of social power relations: above all, those of neoliberal forms of capitalist development and class hegemony. Of particular importance to the approach is Harvey's (1996:196) argument that "much of what happens in the environment today is highly dependent upon capitalist behaviors, institutions, activities, and power structures".

Marxist UPE has critiqued dominant approaches to environmental justice research and mainstream activism. Although they credit the environmental justice literature with building a rich base of empirical data and helping mobilize political activism, Swyngedouw and Heynen (2003:910) argue that it has tended to operate in terms of a "liberal, and hence, distributional perspective on justice", avoiding radical critiques addressing the hegemony of neoliberal capitalism in the organization of human–nature relationships. They also criticize existing environmental justice literature for its "narrow focus"—often confined to local case studies that do not lend themselves to generalization—in contrast with "the broad theoretical perspective employed by political-ecologists" (Swyngedouw and Heynen 2003:910). They emphasize that the conventional focus on dynamics at local scales—often the place-based community—risks overlooking crucial processes and relations generating environmental inequalities at broader regional, national, and global scales (eg, Gandy 2002). In many ways, their assessment echoes and extends Harvey's (1996:400–401) earlier sympathetic critique of the environmental justice movement, which he urged to "transcend the narrow solidarities and particular affinities shaped in particular places" and to confront the "fundamental underlying processes . . . that generate environmental and social injustices": that is, the asymmetrical power relations embedded in "unrelenting capital accumulation".

Bringing a Marxist UPE approach to the St Regis case would thus turn our attention to injustices other than simple distributions; to the multi-scalar processes and relations that generate these injustices; and in particular to the structuring power of neoliberal capitalism. For instance, we might explain the shift in power to assess risk—not just to industry, but to the liable site owners themselves—by situating it or contextualizing it as part of the broader neoliberalization of environmental regulation (Castree 2008a, 2008b; Heynen et al 2007). This neoliberalization handed the owners of St Regis and other large corporations considerable power over the evaluation and management of public health at Superfund sites. On the other hand, it has also generated contradictions and the possibility of contestation, in part by devolving some of the power to regulate to local communities, including reservations. Indeed, the flows of federal funding associated with this devolution provided the support that made the very delineation of a tribal traditional lifeways "receptor" possible. In both of these shifts— the power to assess risk, the power to contest risk assessments—we find evidence that the broader rescaling of governance under the hegemony of neoliberalism extends both to the field of environmental health risk and to particular sites of contestation over risk.

Within this context, a Marxist UPE approach could pose additional questions about the St Regis risk assessment itself. For instance, how did the interests of the liable company find their way into decisions about sampling and analysis, or distort conclusions about the level of risk to human health? How did the broadly neoliberalized terrain of risk assessment both enable and constrain the Leech Lake Band in its efforts to contest the company's site-specific assessment? In short, using a Marxist UPE approach we could explain the particular injustices and inequalities at St Regis by situating, contextualizing, and historicizing them with respect to past and present dynamics of governance and contestation under capitalism.

Against Synthesis: Separating ANT from UPE

So, again, what might an actor-network approach add to the critical geography of environmental justice, at St Regis or any other case? One possible answer to this question has already emerged within Marxist scholarship, and within Marxist UPE specifically. This answer proposes ANT as a useful supplement to an already rich Marxist literature on uneven environments and the production of "social nature". Readings in this vein cast ANT as a valuable resource to strengthen theory and method in UPE and related critical research programs, albeit one with important limitations. In some cases, this has meant invoking ANT as a body of work that draws our attention to, for instance, urban environments as "hybrid assemblages", part social and part natural (eg, Swyngedouw and Heynen 2003). Haraway's (1997) explicitly Marxist and feminist take on ANT has been of particular influence here. A few scholars have gone beyond this, seeking to develop theoretical syntheses linking a compromised or "weakened" variant of ANT with relational forms of Marxism (Castree 2002; Gareau 2005; Perkins 2007).

Despite this acknowledged influence, Marxists and other critical scholars have generally rejected the remainder of ANT as insufficiently critical, theoretical, and explanatory, and therefore less useful as a "standalone" approach to research on injustice, environmental or otherwise. Swyngedouw and Heynen (2003:902–903), for example, suggest that ANT fails to attend to three crucial phenomena: "the social positioning and political articulation" of action; "the contested making of socionatural networked arrangements, rather than . . . a mere dense description of such networks"; and the embeddedness of urban socio-environmental relations in "discursively scripted and culturally imagined understandings". The first of these claims suggests that while ANT usefully points to nonhuman agency, it neglects uneven relations of social and political power (cf Haraway 1997). By implication, an ANT without the support of Marxist theory is unlikely to shed light on social and environmental inequalities like those we encounter at St Regis. The second claim implies that ANT is not only "merely descriptive", but that it also focuses on networks that are already made, while Marxism focuses on processes: the "contested making" of networks. The third claim suggests that Marxist approaches are more sensitive to cultural and discursive dimensions of the production of urban environments. Basically, the suggestion here and in the various efforts to synthesize the two traditions is that ANT can be useful, but only if it combines its strengths with the theoretical apparatus that Marxism can provide. In contrast, other Marxist critics not only reject attempts at reconciliation and synthesis, but also explicitly or implicitly condemn actor-network approaches as inferior, on the grounds that they refuse materialist abstraction (Rudy 2005) and fail to attend to the historical specificity of actor networks (Fine 2005).

While I do not contend that ANT is invulnerable to critique, I suggest here that Latour's (2004, 2005) recent work casts doubt on the validity of some of these claims. First, it is important to distinguish actor-network conceptions of the *social* from those prevalent in both conventional and critical social theory. In both of the latter, *social* continues to refer to *human* relations, institutions, structures, and systems. In one effort at synthesis, for instance, Perkins (2007:1155) distinguishes between "social forms of labor"—where power relations are purportedly located—and "nonsocial forms of labor", such as "the nonsocial labor of biophysical actants like fungi". Kirsch and Mitchell's

(2004:700) contention that ANT neglects the "social intentionality in turning relationships into things" rests on a similar dualism. In such formulations, even if the social/human and natural/nonhuman are conceptualized as dialectically related, they persist as distinctive ontological realms.

Although there is nothing inherently invalid about such dialectical conceptualizations, they are nonetheless inconsistent with Latour's approach to the social. Latour (2005:7) rejects taking the social as a distinct domain of reality, redefining it instead as "a very peculiar movement of re-association and reassembling": one that is heterogeneous, rather than uniquely human. Consequently, in ANT social relations or intentionalities cannot be invoked to *explain* phenomena; on the contrary, ANT seeks to trace the activity of giving shape to "society" and the "social". In order to begin this tracing, a Latourian actor-network analysis turns not to an established social scientific theoretical framework, but instead to *controversies* and *uncertainties* about what the social is composed of: about the identity of groups; about the number of agents involved in courses of action; about the kind of agency that these agents exercise; about the connections between the facts of natural science and the "rest of society", and finally about the nature of social science itself.[5] As critics have pointed out, such an approach carries with it the problem that analytical decisions must be made without the benefit of a body of substantive social theory to guide them (eg, Fine 2005; Rudy 2005). But for Latour (2005:221), "that's just the point". The absence of substantive social theory in ANT generates the possibility of viewing social theories *themselves* in a new light: as circulating accounts aiming (and often competing) to establish the right composition of society, social intentionality, and social action.

In addition, a number of scholars have critiqued ANT for equating or ontologically "leveling" human and nonhuman agency, or at least eliding important differences between them (eg, Kirsch and Mitchell 2004; Laurier and Philo 1999). However, Latour's argument is not that humans and nonhumans exercise the same kind of agency, and that the two are thus ontologically equivalent. On the contrary, his point is that the question of agency is a "profound uncertainty":

> ...the human–nonhuman pair does not refer us to a distribution of the beings of the pluriverse, but to an uncertainty, to *a profound doubt about the nature of action*, to a whole gamut of positions regarding the trials that make it possible to define an actor. (Latour 2004:73; emphasis in original)

The task of an actor-network analysis is not to resolve this uncertainty by explaining which kind of agency—free will, causal necessity, or something in between—belongs to humans and which to nonhumans, or by declaring human and nonhuman agency to be equals. Again, it is to examine *controversies* and *uncertainties* about how agency is distributed, and to trace how such controversies come to be resolved. It is also to see how far such a tracing can be conducted without recourse to existing theories and concepts of the social.

But what do such controversies have to do with environmental inequalities and injustices? If ANT does not explain inequalities as a consequence of social structures and power relations, how can it address them? To be sure, ANT abstains from drawing on inequalities and differences as sources of *explanation*. Nonetheless, Latour's (2005) recent work places considerable emphasis on accounting for the generation of inequalities and differences. In this process of generation, controversies and uncertainties in accounts of nonhuman agency take center stage:

> ...what is new [in ANT] is that *objects* are suddenly highlighted not only as being full-blown actors, but also *as what explains the contrasted landscape we started with*, the overarching powers of society, the huge asymmetries, the crushing exercise of power. (Latour 2005:72; emphasis added)

Marxist UPE also emphasizes that the physical properties of nonhumans play essential roles in the production of injustices. However, Latour's claim above seems to go further, and could easily appear as fetishistic or deterministic. Is he making the anthropomorphic suggestion that nonhuman objects, rather than human power relations, cause inequalities? Or would an ANT analysis claim that environmental inequalities are "natural"—that is, products of natural laws that scientists of various kinds can register? No: Latour's claim suggests that we find multiple, competing *accounts* of what generates inequalities, hierarchies, and relations of domination. In these accounts, objects—frequently taking the form of Nature, with its laws and necessities—are invoked or "highlighted" as what explain these inequalities. On one end of the spectrum, we find the agency of objects cast in the terms of causal *determinism*: climatic, genetic, technological, and so on. Marxist critiques, similarly, have long emphasized the reifying power of such "naturalizing" accounts of nonhuman agency.

However, on the other end of the spectrum, we find accounts in which nonhumans, objects, or Nature play purely *passive* roles—as tools or as backdrops—and where the dominant forces generating inequalities are framed as social, cultural, political, and economic factors. If at the first pole we find various ways of naturalizing inequality and domination, at the latter pole we find assertions that inequality and domination are completely *un*natural: that is, entirely the result of "social" processes. In between the two poles, we find many other accounts of how nonhuman agencies participate in the production of inequality and domination. Marxist UPE would arguably be one of these, although nonhumans sometimes slip into a passive role:

> ...it is exactly those "natural" metabolisms and transformations that become discursively, politically and economically mobilised and socially appropriated *to produce environments that embody and reflect positions of social power*. Put simply, gravity and photosynthesis are not socially produced, of course. However, their powers are *socially mobilised to serve particular purposes*...(Swyngedouw and Heynen 2003:902; emphasis added)

On the one hand, the passage hints at Marxist theory's broader attentiveness to the power associated with naturalizing social configurations. However, the significance of nonhuman agents here lies in their "social mobilization" to produce environments that benefit elite social interests at the expense of others. Such a formulation not only risks slipping back into a society–nature dualism, but also remains incompatible with ANT by its refusal to subject social explanations of inequalities to the same critique it applies to natural explanations.

In an actor-network account, it makes no sense to claim that nonhuman environments express or reinforce power relations confined to a separate social realm. Again, in ANT the role of nonhumans in the social— redefined as the movement of assembling associations—is a site of uncertainty. Rather than revealing how inequalities cast as natural are in fact the results of social relations (or vice versa), an ANT analysis seeks to register the range of competing accounts of agency in the production of domination: from accounts in which nonhuman agency takes a deterministic form, to those in which nonhumans are simply passive objects of social mobilization, to those which lie somewhere in between.

In short, ANT seeks to abstain from *resolving* the question of how nonhumans participate in the production of domination and inequality. But does this refusal to expose social power relations as the causes of inequality simply confirm that ANT is neither explanatory nor critical? Again, the answer is no, but ANT pursues a path that diverges significantly from Marxist UPE. First, in ANT the term *network* is more ambiguous than many interpretations recognize. Latour (2005:131) emphasizes that the term is a "tool to help describe something, not what is being described". It is not, at least in the first instance, a "thing out there that would have roughly the shape of interconnected points", but rather a property of a kind of analytical *account*: "a string of actions where...all the actors do something and don't just sit there" (Latour 2005:128–129). ANT, in other words, is not about describing already formed "social networks", but about tracing the (contested) assembling of such assemblages without taking the existence of social relations—like capitalism and class—for granted.

Second, and more frequently recognized, Latour challenges the distinction between explanation and description. The assertion above that ANT is "mere dense description of networks" implies that a scientific account should appeal to something *else*—something concealed to the actors, but not to the analyst—in order to settle questions of agency or link causes and effects. This is an important assumption in Marxist UPE. According to Swyngedouw and Heynen (2003:906), "the aim of [Marxist] urban political ecology is to expose the processes that bring about highly uneven urban environments". In support of this, they quote Peet's (1977:6) definition of "radical science" as that which "provides alternative explanations which trace the relationship between 'social problems' at the surface and deep societal causes". Such alternative explanations have recourse to the materialist abstractions of Marxist theory, such as value and labor (Kirsch and Mitchell 2004; Perkins 2007; Rudy 2005). But ANT rejects both the notion that the social theorist has privileged access to "deep societal causes" and the idea that such causes exist. Although this means that ANT must indeed do without the powerful theoretical toolkit available to Marxist and other traditions of social theory, it generates an opening for the analyst to conduct research in a quite different way—in the next section, I will attempt to provide at least a hint of what this might mean.

Third, it is also crucial to emphasize that Latour's actor-network approach goes beyond simply registering the ways that competing accounts frame explanations of the controversies, including the question of nonhuman agency in the production of inequality and domination. The next step of an ANT analysis is to trace the *stabilization* of such controversies, including the establishment of some accounts as hegemonic—and thus capable of defining and structuring unequal and potentially unjust configurations of society and nature.

Recent Marxist UPE work has taken up this task as well. Loftus and Lumsden (2008), for instance, draw on Gramscian concepts to account for how the expansion of an urban water network in South Africa produced a new socio-natural configuration in which hegemonic "worldviews" (first apartheid, and later neoliberalism) could be consolidated. Nonetheless, even in such rich and non-reductionist accounts of nonhuman roles in stabilizing particular social projects, there is still arguably an important limitation. As Barnett (2005:9) contends:

[T]he recurrent feature of the political-economy invocation of hegemony is that it lacks any clear sense of how consent is actually secured, or any convincing account of how hegemonic projects are anchored at the level of everyday life, other than implying that this works by "getting at" people in some way or other.

Although Loftus and Lumsden (2008:114) show convincingly that the extension of a water network had effects at the "level of everyday life", their account stops short of demonstrating how the water, "embodying and expressing capitalist hegemony", actually enabled elite hegemonic projects to be incorporated into individual subjectivities.

ANT provides one way to address the "anchoring" of hegemonic projects, including even neoliberal capitalism. Critiques have suggested that ANT, unless married somehow to Marxism, ultimately must remain silent about capitalism; Gareau (2005:129), for instance, charges that ANT is unable to accept "generalized tendencies in capital to affect associations". Although many (perhaps most) ANT accounts have indeed abstained from discussion of capitalism, this need not be the case (Braun 2005b; Latour 2005). However, instead of explaining unjust socio-ecological landscapes as results of capitalist structures, social conflicts, or hegemonic projects, an ANT account would focus on the circulating *forms* and *standards* that do the work of making the tendencies in capital general (Latour 2005). It aims to trace how forms and standards connect local sites and interactions to equally local centers of power, thereby doing the work of *structuring*: defining and producing the "social context", stabilizing "social relations", and consolidating subjectivities, including those characteristic of capitalism and neoliberalism.[6] At the same time, though, Latour's actor-network approach requires abstaining from the assumption that other ways of associating entities—law, religion, science, and so on—are structured entirely by the generalized tendencies of capitalism. Instead, we must ask how "connectors" like these might contribute in quite distinctive ways to the movement of assembling the social.

Since the two traditions operate using such different conceptions of the social, efforts to fuse Marxist and actor-network approaches—particularly in the study of environmental inequalities—are counterproductive. For instance, Gareau's (2005:127) effort to synthesize Marxism and ANT hinges in part on the proposal that "[e]cological Marxism . . . provides 'ANTers' with a context in which to situate their studies of capitalist networks" (cf Kirsch and Mitchell 2004:700). However, the very purpose of an ANT analysis is to trace how actors—through the circulation of forms and standards—*contextualize each other*, not to do the work of "situating" the actors in an a priori context. While Marxist UPE focuses on social structures and relations that have already been assembled, asking how they produce uneven and potentially unjust urban environmental change, ANT attends to the paper trails and information conduits that make possible the assembling of society, context, and relations of inequality and domination. Neither approach refutes or negates the other; they simply work from entirely different starting points, which lead them to entirely different questions (and answers).

Revisiting the Case Study

To return to the risk assessment at the St Regis Superfund site, then: what distinctive questions about environmental justice and injustice would an actor-network approach open up? Instead of *contextualizing* the risk assessment—identifying the social relations of power that can explain its development or its results—an actor-network approach invites us to ask, first, in what distinctive ways do risk assessment and exposure assessment, as codified and practiced under the oversight of EPA, associate entities? Second, through what forms and standards has EPA transported these associations of entities to St Regis? And third, in what ways might the circulation of these associations of entities to St Regis have generated an environmental injustice?

Two of the key "immutable mobiles" through which EPA has codified and standardized much of the practice of risk assessment at Superfund sites are its guidebook *Risk Assessment Guidance for Superfund* (EPA 1989) and the *Exposure Factors Handbook* (EPA 1997). The *Exposure Factors Handbook* gathers together the results of numerous scientific studies, which in combination assemble a set of statistical human populations and subpopulations: adults, children, pregnant women, recreational anglers, and so on. The *Handbook* defines these populations in terms of a set of quantifiable standard default exposure factors, which can then be plugged into equations to calculate likely exposure to contamination at specific Superfund sites: inhalation rates, ingestion rates, exposure frequencies and durations, body weights, life expectancies, and so on. The *Handbook*'s form—whether print or digital—makes it simple to transport these statistical aggregates, compiled and translated from a vast multiplicity of sites and sources, to contaminated sites like St Regis. At St Regis and other Superfund sites the exposure factors are combined with other circulating default factors (including assumptions about toxicity, gathered from yet another vast array of laboratory rodent assays and epidemiological studies) to calculate estimates of risk to human health.

The question that has arisen at St Regis and other sites that affect American Indian reservations and tribal lands is whether these circulating exposure factors do justice to tribal traditional lifeways (Harris and Harper 1997; Minnesota Sea Grant and Natural Resources Research Institute 2002). First, are individuals who practice such

lifeways likely to be exposed to higher levels of contamination
than the hypothetical individuals assembled in the *Exposure Factors
Handbook*—for instance, by eating larger amounts of fish or consuming
wild plants? Although the most recent edition of the *Handbook*
provides alternative estimates of fish ingestion for a generalized "Native
American subsistence population", it does not register differences
among or within tribes, or provide estimates for other exposure
pathways. Second, is the hypothetical human *individual* the appropriate
"social unit" at risk, or should risk assessment evaluate risk to another
unit, such as the tribal community or culture—which might be threatened
with extinction if forced to eliminate traditional practices (Harris and
Harper 1997)? Or, more radically, the wider webs of human–nonhuman
relations that constitute traditional American Indian worlds? In light
of the latter, an environmental justice human health risk assessment
panel report for the St Regis site suggested that EPA should abandon its
conventional practice of ordering separate human health and ecological
risk assessments (Minnesota Sea Grant and Natural Resources Research
Institute 2002). And third, is there any way for circulating standards for
exposure and risk assessment to take into account the less tangible, less
quantifiable aspects of tribal traditional lifeways, such as the spiritual
attachment of a people to a particular land and environment (National
EPA-Tribal Science Council 2006)?

When the Leech Lake Band of Ojibwe set out to develop its
own tribally specific traditional lifeways exposure scenario for the
St Regis human health risk assessment (Leech Lake Division of
Resource Management 2003), it sought to rectify a distinctive kind
of environmental injustice: the invisibility of tribal traditional lifeways
in EPA's approach to risk assessment, the failure of this approach to
take them into account and register their presence. Although such an
injustice involves distributive and procedural aspects, it extends beyond
these. The problem is not simply that exposure to the contamination
at St Regis may have been distributed unequally, or that the process
of assessing and managing health risk failed to give traditional tribal
members a voice, but that risk assessment as a way of associating entities
cannot fully register the *difference* of tribal traditional lifeways—itself
yet another set of relations, another way of associating entities—in
and around Cass Lake. And so the pursuit of environmental justice
encompassed not only the efforts to get the site investigated and cleaned
up, but also to make visible tribal traditional lifeways as meaningfully
different associations of humans and nonhumans.

My point is not that a Marxist approach is incapable of explaining
such an injustice, by situating it within a social context, although I will
leave it to others to venture such an explanation. It is simply that an actor-
network approach, by looking first not to processes in an established

ontological domain of the social, but instead to risk assessment as a way of associating—and a means of circulating entities via forms and standards to make them general—opens up distinctive questions and requires different ways to go about answering them. Rather than asking which unequal power relations risk assessment might reflect or instantiate, we have to focus first on what risk assessment does and how it works: the particular way it gathers, associates, and circulates statistical individuals and populations at risk. The shift in focus might then enable us to see if and when risk assessment changes from little more than an *intermediary*, expressing particular economic interests, to a *mediator* mobilizing non-humans in various ways to make a difference in the process of assembling the social.

The Political Usefulness of ANT

Still, the question remains: if it remains unwilling to "strip away diversions" and unveil hidden social forces, can ANT be *critical*? Or is it indeed "politically useless" as charged (Rudy 2005)? In order to be relevant to the political project of environmental justice, ANT obviously cannot "simply enjoy the spectacle of the sheer multiplicity of new connections" (Latour 2005:259).

Latour (2005:261) characterizes the political intervention of an actor-network account in the following terms: "simply to highlight the stabilizing mechanisms so that the premature transformation of matters of concern into matters of fact is counteracted". In environmental justice research specifically, actor-network approaches can be critical by producing a representation of the conduits and means through which potentially unjust ways of associating people and things become dominant or hegemonic at the expense of others. In the St Regis case, for instance, the task was to highlight the forms and standards through which risk assessment circulates the entities it associates, and to trace the specific openings for contestation within it, such as assumptions about exposure through the practice of tribal traditional lifeways (cf Law and Mol 2008). This is quite different from revealing the social interests hidden behind risk assessment's ostensibly neutral assumptions, processes, and outcomes. In other cases, an ANT account might well play a small role in combating the rush to naturalize inequalities and domination—that is, to institute some social forms, like neoliberalism, as laws and essences to which there is no alternative.

Nonetheless, it is true that an ANT account does not *automatically* have political relevance. It certainly can (and perhaps usually does) remain as "politically useless" as any other academic account risks becoming. An actor-network account should be an effort to "help define the right procedures for the composition of the collective *by rendering itself interesting* to those who have been the object of study" (Latour 2005:160; emphasis added). In other words, the test of its political relevance and usefulness should be the same as that of any other social science account: whether it is taken up by those it has sought to represent, and whether it enters into the ongoing composition of society. At its best, an ANT account is a "risky account" that stands a good chance of failing; for Latour, it is the fifth source of uncertainty.

What does it mean to counteract the "premature transformation of matters of concern into matters of fact"? It is still a common belief that we can settle controversies—including uncertainties about environmental inequalities—by appealing to facts provided by social and natural scientists. But a growing number recognize now that scientists can no longer plausibly settle disputes simply by providing the "hard facts". We live in a time when one kind of scientific certainty about nature—and perhaps most prominently, certainty about the state of the environment and its connection with human health and wellbeing—has largely disappeared (eg, Beck 1992; Bingham and Hinchliffe 2008; Giddens 1984). It is no longer possible to establish "a common front of indubitable matters of fact that politicians could subsequently use in support of their decisions" (Latour 2004:63). Few things illustrate this better than the phenomenon of "dueling experts", which frequently figures in conflicts over environmental health and justice (Nelkin 1995). When the facts presented by scientific experts contradict each other, we cannot appeal to them to settle disputes; we cannot turn to Nature to resolve our arguments. In the face of this multiplication of controversies, Latour (2004:63) argues that we can take one of two attitudes: we can wait for the sciences to overcome uncertainties, by generating more and better data and improved methods, or we can "consider uncertainty as *the inevitable ingredient of crises in the environment and in public health*" (emphasis added). In such crises, we do not have indisputable matters of natural or social fact; we have instead highly disputable matters of concern.

How does ANT's effort to counteract the premature institution of facts relate to the political agenda of Marxist UPE? Swyngedouw and Heynen (2003:898) describe this agenda in clear terms:

> ...to enhance the democratic content of socioenvironmental construction by identifying the strategies through which a more equitable distribution of social power and a more inclusive mode of environmental production can be achieved.

They go on to characterize an "emancipatory urban politics" as, at least in part:

> ...acquiring the power to produce urban environments in line with the aspirations, needs and desires of those inhabiting these spaces—the

capacity to produce socially the physical and social environments in which one dwells. (2003:915)

Although such a political project is commendable, there are key points of tension between it and an actor-network approach. First, enhancing "the democratic content of socioenvironmental construction" appears to mean extending meaningful participation in decision-making processes to a wider range of *human* actors—in particular, people who have been marginalized by capitalism, racism, sexism, colonialism, and so on. Again, the distribution of "social power" and the concept of "social production" evidently refer to a human domain. The agenda evokes the familiar concept of procedural environmental justice, but here tied to a radical critique of capitalism.

Although Latour's conception of democracy is also tied to process, it is not limited to the deliberations and demonstrations of human actors. However, to extend democracy to nonhumans is not to invite trees and chemicals to public meetings and demonstrations. Instead, it is to recognize that nonhumans are already involved in democracy, through the competing accounts that mobilize them—often in the determinist form of Nature—and through the disputable matters of concern at the center of so many controversies. In the case of a Superfund risk assessment, this means that democracy and environmental justice require attending not just to who participates and how, but also—for example—to questions about how to sample and analyze soil or fish, or what to extrapolate from a series of lab rodent bioassays. In this sense, enhancing democracy means not only extending the range of *human* participants in political decision-making, but also taking full account of nonhuman participation in the assembling of the social.

A second point of tension pertains to the political relevance of critical research. The Marxist UPE agenda cited above offers to "identify the strategies" for achieving greater equality and inclusiveness in the production of urban environmental change, in part by stripping away diversions and unearthing deep societal causes of socio-environmental injustice. Latour argues that such an approach, rather than supporting an emancipatory politics, in fact both forecloses the possibility of empirical surprise and short-circuits the political process. By demonstrating repeatedly how environmental injustices are generated by the social relations and dynamics of advanced capitalism, Marxist UPE risks "becoming empirically empty and politically moot" (Latour 2005: 251):

> When faced with new situations and new objects, it risks simply repeating that they are woven out of the same tiny repertoire of already recognized forces: power, domination, exploitation, legitimization, fetishization, reification. (Latour 2005:249)

This is, in effect, a reiteration of an older argument that critical approaches of this kind "can never fail to be right" (Vayda and Walters 1999). In addition, stepping in to resolve controversies with powerful explanations that appeal to hidden social causes risks depoliticizing the controversies every bit as much as neoliberalism. Marxist UPE indeed challenges neoliberalism's naturalization of particular social formations. But to the extent that it does so by revealing the deep societal causes of inequality, it risks presenting us with an alternative set of indisputable *social* facts. Latour outlines a different way in which ANT could enhance democracy: by slowing down the process of resolving controversies about what the world is made of.

Latour is frequently harsh—and perhaps unfair—in his treatment of critical social theories, but he does not dismiss them either. First, he contends that critical approaches like Marxism are crucial precisely because they raise the question of political relevance in social research. ANT aims to respect and preserve this critical impulse. Indeed, for Latour (2005:256) it is positivism that is "wrong politically", because it rests on a false dichotomy between disinterested science and engaged politics. Second, although Marxist explanations should not be taken as indisputable resolutions of the question of what causes environmental injustices, they are nonetheless important because they prepare us for the political task of composing a just common world. This is why an actor-network approach to environmental justice needs Marxist UPE: not as a partner in an awkward and unwieldy theoretical synthesis, but instead as a rich source of contestable narratives, hypotheses, arguments, models, and explanations for the production of environmental injustices. ANT's critical role is not to verify or refute such explanations—or those of neoclassical economics, for that matter—but to trace their participation in the assembling of the social. For example, how do some accounts of environmental injustice, and not others, prevail and become instituted as "matters of fact"?

Marxists and actor-network theorists can debate whether the democratization of socio-environmental production is better served by revealing hidden causes or by tracing the movements that connect localized, structuring centers of power with the equally local interactions and conflicts that they format. And perhaps a critical UPE approach to environmental justice can benefit from some of both. Ultimately, the test of political relevance for either approach will be the wider democratic arena outside the narrow circuits of critical academic geography.

Conclusion

My primary contention, in brief, is that an actor-network approach opens up lines of inquiry in critical environmental justice research that Marxist UPE does not. Early empiricist studies of environmental inequalities brought academic attention to uneven distributions of environmental risks and to temporal changes in these patterns. Marxist UPE has expanded our conception of what "environmental inequality" encompasses, challenged us with the question of justice at different scales, and provided theoretically grounded explanations of environmental injustices. ANT now has the potential to generate a new body of environmental justice research: one that, instead of describing or explaining environmental inequalities, traces the emergence and resolution of controversies and uncertainties surrounding environmental injustices.

The way forward is neither to thank ANT for alerting us to nonhuman agency (while dismissing it for its alleged inattention to power relations) nor to build syntheses that attempt to reconcile ANT with Marxism. Such efforts both miss the distinctive contributions that ANT can make and gloss over its incommensurability with other social-scientific traditions of explanation. Nor is it to counter arguments that Marxist approaches are superior to ANT with claims that the reverse is true. ANT does not replace other critical approaches to environmental injustice, but provides an alternative way in. Let us simply *add* ANT to the repertoire of approaches to environmental justice research.

But let us add it to the repertoire of *critical* approaches to environmental justice research as well. The aim of an actor-network analysis should not be to stand back from the political fray and generate neutral descriptions featuring gratuitous hordes of nonhumans. On the contrary, it should be to participate in a modest way in the re-composition of the common world. Part of this participation consists in "de-composition"—not *de*construction, but *re*construction of the tiny conduits through which big structuring forces like capitalism and neoliberalism, or distinctive ways of associating like risk assessment, are given shape, stabilized, and made hegemonic. But a crucial aim of this reconstruction is to register new and surprising associations, and to experiment with possibilities for reassembling them in better, more just ways—because the new associations are not always good ones. Instead of stripping away diversions, ANT attends closely to them, seeking to trace—without appealing to hidden forces—the means through which they frequently gain stability and hegemony.

Unless we restrict the meaning of "radical" to the uprooting of totalizing systems, an actor-network approach can also, perhaps, even be radical. ANT is about composing a better, more just common world, and it does not shy away from waging battles to make this happen. But it steers the combat away from the totalizing, overwhelming systems, towards the tiny, fragile conduits that give them unity and stability:

Is it not obvious then that only a skein of weak ties, of constructed, artificial, assignable, accountable, and surprising connections is the only way to begin contemplating any kind of fight? (Latour 2005: 252)

Perhaps ANT has weapons it can bring to a critical, radical politics of environmental justice after all.

Acknowledgements

The author thanks Kristin Sziarto and three anonymous referees for their extremely helpful and insightful critiques and suggestions. The empirical research cited in this chapter was supported by a STAR Graduate Research Fellowship from the US EPA and a Consortium on Law and Values in Health, Environment & the Life Sciences Grant, Doctoral Dissertation Fellowship, and Mark and Judy Yudof Fellowship from the University of Minnesota.

Endnotes

[1] In addition to this volume, in the last 2 or 3 years there have been special issues on environmental justice in *Geoforum*, *Society and Natural Resources*, *Health and Place*, and the new journal *Environmental Research Letters*, just to name a few.

[2] For just a few examples of this rich and growing literature, see, eg, Heynen (2003, 2006), Heynen, Perkins and Roy (2006), Njeru (2006) and Veron (2006).

[3] Remarkably, ANT garnered this reputation even as conventional scientists dumped its investigations of science in the same category as postmodernism, deconstruction, and "stronger" forms of social constructionism, all of which they condemned for politicizing and misinterpreting the sciences (Sokal and Bricmont 1998).

[4] Although Latour does not represent all "actor-network theorists", I focus on his interpretation here since he is the only one who has attempted to lay out ANT as a relatively systematic approach. I place particular emphasis on Swyngedouw and Heynen's agenda for Marxist urban political ecology for the same reason.

[5] Rudy (2005:112) charges that Latour's ANT fails in part because it rejects dualism while it pursues "the description of a metaphoric middle ground that doesn't and can't exist without the poles [of, eg, Society and Nature]." But ANT rejects the nature–society dualism primarily as a *starting point* for analysis and for politics. If a key component of modernity is the division of reality into nature and society, a key task of ANT is to analyze the work that produces, maintains, and realigns this division.

[6] Latour (2005) calls these centers of power—of which his famous "centers of calculation" are but one variety—*oligoptica*.

References

Barnett C (2005) The consolations of "neoliberalism". *Geoforum* 36:7–12
Beck U (1992) *Risk Society: Towards a New Modernity*. London: Sage Publications
Bingham N and Hinchliffe S (2008) Reconstituting natures: Articulating other modes of living together. *Geoforum* 39:83–87
Braun B (2005a) Environmental issues: Writing a more-than-human urban geography. *Progress in Human Geography* 29:635–650
Braun B (2005b) Writing geographies of hope. *Antipode* 37:834–841

Buzzelli M and Veenstra G (2007) New approaches to researching environmental justice: Combining critical theory, population health and geographical information science (GIS). *Health and Place* 13:1–2

Castree N (2002) False antitheses? Marxism, nature and actor-networks. *Antipode* 34:111–146

Castree N (2008a) Neoliberalising nature: The logics of deregulation and reregulation. *Environment and Planning A* 40:131–152

Castree N (2008b) Neoliberalising nature: Processes, effects, and evaluations. *Environment and Planning A* 40:153–171

EPA (1989) *Risk Assessment Guidance for Superfund*. Washington DC: Office of Emergency and Remedial Response, US Environmental Protection Agency

EPA (1996) *Revised Policy on Performance of Risk Assessments During Remedial Investigation/Feasibility Studies (RI/FS) Conducted by Potentially Responsible Parties*. Washington DC: Office of Solid Waste and Emergency Response

EPA (1997) *Exposure Factors Handbook*. Washington DC: US Environmental Protection Agency

Fine B (2005) From actor-network theory to political economy. *Capitalism Nature Socialism* 16:91–108

Gandy M (2002) *Concrete and Clay: Reworking Nature in New York City*. Cambridge, MA and London: MIT Press

Gareau, B J (2005) We have never been human: Agential nature, ANT, and Marxist political ecology. *Capitalism Nature Socialism* 16:127–140

Giddens A (1984) *The Constitution of Society: Outline of the Theory of Structuration*. Berkeley and Los Angeles, CA: University of California Press

Haraway D J (1997) *Modest Witness@Second Millennium. FemaleMan Meets OncoMouse: Feminism and Technoscience*. New York: Routledge

Harris S G and Harper B L (1997) A Native American exposure scenario. *Risk Analysis* 17:789–795

Hartwick E R (2000) Towards a geographical politics of consumption. *Environment and Planning A* 32:1177–1192

Harvey D (1996) *Justice, Nature, and the Geography of Difference*. Oxford: Blackwell Publishers

Heynen N (2003) The scalar production of injustice within the urban forest. *Antipode* 35:980–998

Heynen N (2006) Green urban political ecologies: Toward a better understanding of inner-city environmental change. *Environment and Planning A* 38:499–516

Heynen N, McCarthy J, Prudham S and Robbins P (eds) (2007) *Neoliberal Environments: False Promises and Unnatural Consequences*. London and New York: Routledge

Heynen N, Perkins H A and Roy P (2006) The political ecology of uneven urban green space: The impact of political economy on race and ethnicity in producing environmental inequality in Milwaukee. *Urban Affairs Review* 42:3–25

Holifield, R (2007) "Spaces of risk, spaces of difference: Environmental justice and science in Indian country." Unpublished PhD thesis, University of Minnesota

Keil R (2003) Urban political ecology. *Urban Geography* 24:723–738

Kirsch S and Mitchell D (2004) The nature of things: Dead labor, nonhuman actors, and the persistence of Marxism. *Antipode* 36:687–705

Kurtz H (2003) Scale frames and counter scale frames: Constructing the social grievance of environmental injustice. *Political Geography* 22:887–916

Laurier E and Philo C (1999) X-morphising: Review essay of Bruno Latour's *Aramis, or the Love of Technology. Environment and Planning A* 31:1047–1071

Latour B (2004) *Politics of Nature: How to Bring the Sciences into Democracy*. Cambridge, MA: Harvard University Press

Latour B (2005) *Reassembling the Social: An Introduction to Actor-Network-Theory*. Oxford and New York: Oxford University Press

Law J and Mol A (2008) Globalisation in practice: On the politics of boiling pigswill. *Geoforum* 29:133–143

Leech Lake Division of Resource Management (2003) *Leech Lake Band of Ojibwe Pilot Superfund Project, Final Report*. Cass Lake, MN: LLDRM, Environmental Quality Services

Loftus A and Lumsden F (2008) Reworking hegemony in the urban waterscape. *Transactions of the Institute of British Geographers* 33:109–126

McCarthy J (2002) First World political ecology: Lessons from the Wise Use Movement. *Environment and Planning A* 34:1281–1302

Minnesota Sea Grant and Natural Resources Research Institute (2002) *Assessing and Communicating Risk: A Partnership to Evaluate a Superfund Site on Leech Lake Tribal Land: Final Report to U.S. EPA, Environmental Justice Program*, Grant No EQ825741

National EPA-Tribal Science Council (2006) Paper on tribal issues related to tribal traditional lifeways, risk assessment, and health & well being: Documenting what we've heard. US Environmental Protection Agency, April 2006

Nelkin D (1995) Science controversies: The dynamics of public disputes in the United States. In S Jasanoff, G E Markle, J C Petersen and T Pinch (eds) *Handbook of Science and Technology Studies* (pp 444–456). Thousand Oaks, CA: Sage Publications

Njeru J (2006) The urban political ecology of plastic bag waste problem in Nairobi, Kenya. *Geoforum* 37:1046–1058

Peet R (1977) The development of radical geography in the United States. In R Peet (ed) *Radical Geography: Alternative Viewpoints on Contemporary Social Issues* (pp 6–30). Chicago: Maaroufa Press

Perkins H A (2007) Ecologies of actor-networks and (non)social labor within the urban political economies of nature. *Geoforum* 38:1152–1162

Powell M R (1999) *Science at EPA: Information in the Regulatory Process*. Washington DC: Resources for the Future

Pulido L (2000) Rethinking environmental racism: White privilege and urban development in southern California. *Annals of the Association of American Geographers* 90:12–40

Rudy A P (2005) On ANT and relational materialism. *Capitalism Nature Socialism* 16:109–125

Schlosberg D (1999) *Environmental Justice and the New Pluralism: The Challenge of Difference for Environmentalism*. Oxford and New York: Oxford University Press

Schroeder R A (2005) Debating the place of political ecology in the First World. *Environment and Planning A* 37:1045–1048

Schroeder R A, St Martin K, Wilson B and Sen D (2008) Third World environmental justice. *Society and Natural Resources* 21:547–555

Schweitzer L and Stephenson M J (2007) Right answers, wrong questions: Environmental justice as urban research. *Urban Studies* 44:319–337

Sokal A and Bricmont J (1998) *Fashionable Nonsense: Postmodern Intellectuals' Abuse of Science*. New York: Picador

Stephens C (2007) Environmental justice: A critical issue for all environmental scientists everywhere. *Environmental Research Letters* 2:np

Swyngedouw E (2004) *Social Power and the Urbanization of Water*. Oxford: Oxford University Press

Swyngedouw E and Heynen N C (2003) Urban political ecology, justice and the politics of scale. *Antipode* 35:898–918

Sze J and London J K (2008) Environmental justice at the crossroads. *Sociology Compass* 2:1331–1354

Taylor D (2000) The rise of the environmental justice paradigm: Injustice framing and the social construction of environmental discourses. *American Behavioral Scientist* 43:508–580

Vayda P and Walters B B (1999) Against political ecology. *Human Ecology* 27:167–179

Veron R (2006) Remaking urban environments: The political ecology of air pollution in Delhi. *Environment and Planning A* 38:2093–2109

Walker G P and Bulkeley H (2006) Geographies of environmental justice. *Geoforum* 37:655–659

Chapter 3

Gendered Geographies of Environmental Injustice

Susan Buckingham and Rakibe Kulcur

Introduction

Environmental justice has been gaining traction as it evolves from an advocacy practice (which it still is) to both a subject for intellectual research and an analytical approach to environmental inequalities. Moreover, the call for environmental justice has been taken up by a number of environmental organizations, particularly in Europe (see, for example, Friends of the Earth and Greenpeace). While the theoretical and methodological frameworks used by academics trying to understand and explain environmental injustice are varied, they tend to focus on inequalities in income and/or race,[1] while neglecting those of gender. Both the US Environmental Protection Agency's definition ("Environmental justice is the fair treatment for people of all races, cultures and incomes regarding the development of environmental laws, regulations and policies") and that provided by Capacity Global for the UK "as equal access to a clean environment and equal protection from possible environmental harm irrespective of race, income, class or any other differentiating feature of socio-economic status" (Schwarte and Adebowale 2007:13) focus on race/ethnicity and class/income as defining features of environmental injustice. Its origins in the predominantly African-American rural hazardous waste site protests in the 1980s, which focused on the injustice of environmental negativities being clustered in minority ethnic communities (Bullard 1999), has focused attention on race and ethnicity as both an explanation for, and mobilizing tool against, environmental injustice at the community level: the geographical scale at which the environmental impact is most strongly perceived. More recently, and particularly in the UK, relative poverty has been acknowledged as a dimension of environmental injustice (Agyeman 2000; Dunnion and Scandrett 2003; Friends of the Earth Scotland 2000; Walker et al 2005).[2] However, the relationship between environmental injustice and other forms of social disadvantage delineated by age, disability and gender remains poorly articulated by all but a handful of commentators and activists (although for an exception, see the London Sustainability Exchange 2004).

We argue in this chapter that environmental justice, as a campaign focus, and as an analytical framework (along with other normative theories of poverty, inequality, wellbeing, social injustice and policy reform, on which see Robeyns 2008), needs to broaden its focus to address a wider range of inequalities than it currently does. Arguing that class has been neglected as a factor structuring environmental injustice, Pellow and Brulle have called for a wider consideration of inequality on the basis that "Scholars cannot understand—and policy makers cannot prevent—environmental injustices through a singularly focused framework that emphasizes one form of inequality to the exclusion of others". They continue to make a case that "environmental injustices affect human beings unequally along the lines of race, gender, class and nation, so an emphasis on any one of these will dilute the explanatory power of any analytical approach" (Pellow and Brulle 2005:298). While it is encouraging to note such a call in a literature largely devoid of references to gender, it has to be noted that it is made in a book in which this is virtually the only mention of gender.

By focusing on gender inequality as a component of environmental injustice we consider the usefulness of concepts of scale and intersectionality as ways of widening consideration of environmental injustice. As Pellow and Brulle have argued, environmental injustice needs to be regarded within the larger "social dynamics of the social production of inequality and environmental degradation" (Pellow and Brulle 2005:3). No society is one-dimensionally exclusive: a society which is racist and classist is likely also to be sexist and to marginalize others by age, disability or frailty. It is not enough, then, for any analysis to focus on a limited number of components of environmental injustice without an acknowledgement of wider structures of power and prejudice. Krall's paraphrase of Blatchford on the importance of adopting the perspective of native people as a guide to nature cautions: "Locked in the mentality of politics, they ('persons who have lost touch with Mother Earth') seek the greatest good for the greatest, but they should not move until the least is cared for" (Krall 1994:81).

In part, we would argue, this neglect of gender is due to the less publicly visible scales at which the consequences of environmental injustice for women, less geographically concentrated than ethnic minorities or people in poverty, are manifested (although, as Bradshaw et al 2003 identify for the UK, women outnumber men in poor communities): the individual body and the household or family. More combatively, we also suggest that it is the organizational structures of the campaigning groups, mobilizing around environmental and environmental justice issues, which contribute to this neglect. Large environmental NGOs, increasingly drawn into advising and lobbying institutions on environmental matters, including environmental justice, together with the institutions themselves, draw the contours of what does, and does not, constitute environmental injustice (see also Brulle and Essoka 2005). While organizations combating environmental injustice deserve praise and support for raising racism, as well as low income, as factors determining unfairly distributed environmental benefits and problems, we argue that they now need to widen their concerns to embrace other forms of inequality if real environmental justice has a hope of being achieved. Our focus in this chapter, based on our own research and environmental activism, is gender. We argue this on the basis that a particularly powerful alignment of factors conspires to marginalize gender: lack of visibility of environmental gender injustice; campaigning organizations which are themselves gender blind at best, masculinist, at worst; institutions at a range of scales which are still structured by gender (as well as class and race) inequalities; and an intellectual academy which continues to marginalize the study of gender—and women's—inequality.

The chapter will first argue that the lack of a gender perspective in most of the environment injustice literature limits both the analyses themselves as well as opportunities for improving environmental justice for other groups. Following this, constructions of scale are explored to see how these might constrain a gender-fair approach to environmental justice. Using an example of waste management to illustrate how gendered institutional structures, and a failure to interrogate inequality within the household, compound environmental injustice, the chapter will examine the success or otherwise of "gender mainstreaming" policies, against a background of continued gender inequalities in environmental organizations and other institutions.

Women and the Environmental Justice Literature

Analyses of environmental injustice have tended to revolve around the nature of justice (Schlosberg 2007; Walker and Bulkeley 2006), the contours of environmental justice and injustice, both generally and through case studies (Agyeman 2005; Checker 2005; Pellow and Brulle 2005; Pulido 1996), and its tendency to exclude broader issues of sustainability (Agyeman, Bullard and Evans 2003). Leaving aside the few explicitly focused articles on gender and environmental justice such as those by Kurtz (2007), Nightingale (2006) and Sze (2004, 2006), and, often tokenistic, chapters in books on environmental injustice which elsewhere fail to engage with gender, these analyses, as the movements themselves, have mostly ignored the "gender question". Key recent books on environmental justice have conspicuously failed to develop an analysis which includes women and gender, while at the same time raising issues which ecological feminist arguments could illuminate. For example, Schlosberg (2007) argues that environmental injustice analyses fail to acknowledge ecological justice, although this link has been made by ecofeminists seeking to develop an ethic of care, which embraces non-human as well as human life (Merchant 1996, 2005; see also Plumwood for a "realign[ment of] reason . . . with social formations built on radical democracy, co-operation and mutuality" (1993:196). Taylor's claim that environmental justice is "the first paradigm to link environment and race, class, gender, and social justice concerns in an explicit framework" (Taylor 2000:542) likewise neglects the work done by ecofeminists in the past, although this is belied by her earlier, innovative contribution on women, race and environmental justice to Karen Warren's collection of writings on ecofeminism (1997). Nor is her claim borne out by analyses of environmental injustice, despite the majority of grassroots protests being generated by women, and in the USA, mostly women from low-income communities, and women of colour (Di Chiro 1998). That theories of environmental justice have generally failed to take gender into account is not surprising. According to Robeyns, "most justice theories do not make their account of gender relations explicit and do not respond to feminist critiques, but are implicitly relying on androcentric and gender-biased assumptions" (2008:89). However, the dialogue between eco-socialists and ecofeminists, which has addressed the nexus between class and gender through structural causes of inequality concerning socio-environmental relationships, suggests scope for this engagement to be taken up by those investigating environmental injustice (see Godfrey 2006; Goldstein 2006; Salleh 2006; Turner and Brownhill 2006, all in a special issue of *Capital, Nature, Socialism*).

There are a number of reasons which may explain this hitherto missed opportunity. For example, MacGregor (2006) argues that ecofeminism's focus on care and the private realm reduces its potency for an environmental justice grounded in citizenship and public space, while a number of writers have observed a machismo prevalent in the environmental movement which works against women, and women's interests (MacGregor 2006; Salleh 2008; Seager 1993). Shrader-Frechette has also noted sexism (as well as racism) in the media, which she believes fails to give voice to the grassroots environmental justice movement because it is "led largely by women of color" (Shrader-Frechette 2002:13). This highlights the importance of considering intersectionality, through which the overlap between vectors of disadvantage is revealed more clearly.

MacGregor (2006) makes a powerful challenge to what she sees as the reification of care work as women's, although we would argue that women's continuing prominence in both low-paid and unpaid caring demands that it be a consideration of policy as well as campaigning. A good example of this can be found in McDowell's work on an "ethic of care" in which she invokes the "relational view of self", arguing that it is "undeniable that there are clear associations between gender, caring and philosophical attitudes based on common practices and responsibilities" (McDowell 2004:156), which are not well rewarded in contemporary, neoliberal society. Likewise, Robeyns (2008), in critiquing social theories of wellbeing and justice from a capabilities perspective,[3] argues that these frequently neglect the, often invisible, care responsibilities undertaken by women. It is also useful to consider the work of Iris Marion Young in this context, whose arguments concerning the importance of procedural justice have influenced some writing on environmental justice (for example, Schlosberg 2007, and see later in this chapter). What these writers have not taken up, however, is Young's analysis of the division of labour as an institutional practice, which has a bearing on justice. In particular, she considers the exploitation of women in the home and in the workplace, through their provision of care, a point this chapter will take up later.

Feminist discussion of "intersectionality", which "locates gender within contexts of race, ethnicity, class and citizenship among other socially constructed categories that interact to shape women's lives and social identities" (Lee 2007:382; and see Cuomo 1994), also offers environmental justice analyses a potentially useful way into considering gender. Carolyn Merchant (1996, 2005) reworked her own earlier positions on ecofeminism in favour of what she has called a "partnership ethic", which "...treats humans (including male partners and female partners) as equals in personal, household and political relations and humans as equal partners with (rather than controlled by or dominant over) nonhuman nature" (Merchant 2005:196).

Scale

Through increasingly nuanced understandings of the concept of scale (Marston 2000), it is apparent that scales are both relational and imbricated in complex ways, as well as being structured by internal and external politics of scale (Brenner 2001). What manifests locally, within the household, or even on the body, is as likely as not to have its genesis at the wider scales, which more commonly come under the purview of economists, political scientists, and mainstream geographers.

It has been argued elsewhere that environmental injustices specifically affecting women often operate at scales which seem invisible to those not actively studying and searching for them (Buckingham 2006, 2007). This is illustrated by Brenner arguing that the household has only an "internal politics of scale", rendering it indistinguishable from a "place" or a "locale" (Brenner 2001:596). This fails to recognize cross-scale politics which tie the household with various levels of governance, political decisions and business practice, and illustrates the invisibility of the domestic scale to many academics. One need only consider the burden placed on households to adopt environmental strategies such as recycling or energy efficiency to "save the planet"—actions which normally fall to women in the household (Buckingham, Reeves and Batchelor 2005; Defra 2002; MacGregor 2006)—or the enthusiasm business has shown for flexible working at home to offset its own operational costs, to recognize how the household is deeply entwined with other scales.

Marston and Smith, in their rejoinder to Brenner, argue that the "scale of the household (and the home) is integral to [scaled social] process[es]" (Marston and Smith 2001:616). A case they also make for considering the body as a valid scale. Mostly, however, the argument that the household and the body are as valid as any other scale is made by feminist geographers such as Linda McDowell (1999). Young proposes that feminist movements were in the "vanguard" of insurgency movements, which began as collectives designed as much to empower women as to politicize them. Such insurgency movements, she writes, exploit and expand the space between individuals and families on the one hand, and states and large corporations on the other (Young 1990:82–85), points illustrated by the relatively rare accounts of gendered resistance to environmental injustice (see, for example, Gibbs 1996; Kurtz 2007; Sze 2004).

The Body

Another argument for environmental justice activists and commentators to broaden their conceptual framework to include a gender analysis is the feminist critique of society "othering" the female as a variant of the "normal" male. This questions the validity of testing chemicals and pharmaceuticals on fit male bodies, when it is known that women's bodies (and children's, the elderly, frail and other groups similarly constituted as "different") will experience these products in different ways (EEA 2003).

A point conceded by the European Union, though not entirely dealt with in its REACH[4] legislation, is that pregnant and nursing women, as well as women who plan to bear children in the future, are particularly vulnerable to environmental pollution. This links to environmental justice in that as well as combating pollution at source, or campaigning for this to be redistributed more fairly, or redistributed away from those most at risk (which may well result in a very different distribution of polluting industries), there should be a much more fundamental argument for challenging any substance release that cannot be proved to be harmless to the most vulnerable. At present, as long as a substance can be expected not to harm the most robust, who, by definition, are the ones tested, it is deemed acceptable. Although this precautionary approach to environmental pollution is technically recognized by a number of international organizations (for example, through European Union law), it is a long way from thorough implementation.

Sze (2006), uses the example of DES (diethylstilbestrol—an artificial oestrogen prescribed to women in the 1940s and 1950s to prevent miscarriages, treat menopause, and other hormonal "problems", as well as to fatten up livestock) to illustrate the gendered nature of its ultimate ban. Although its carcinogenic effect on females generated a challenge to the US Federal Drug Administration by the women's health movement, Sze explains that a ban was only applied once the unnatural—that is feminized—"sexual and bodily development" was noted in men who had eaten DES-fed poultry (Sze 2006:806). The less visible, though obviously harmful, impacts on women were not similarly acted upon "until there was a greater body of scientific evidence" (Sze 2006:797). Such readings, she argues, have led to an emerging recent literature of environmental injustice arising from literary and cultural studies, to culturally redefine "the environmental to mean...human bodies, especially in racialized communities, in cities, and through labor" (Sze 2006:792).

Most ecofeminist analyses argue that masculine identity "depends on men distancing themselves from the fact [of being part of nature]" (Salleh 1997:13). This is not to argue that women are essentially, qua women, closer to nature, but are confronted with nature more frequently and directly by the unpaid and low-paid work they do. Similarly, we would argue that, as with the social model of disability (Shakespeare and Watson 2002), it is the continued structuring of females as "different" to a male norm which results in a biological manifestation of a socio-environmental problem. One example is that of dioxins which bio-accumulate and which dissolve in fats, making them particularly pernicious chemicals for women who have a greater body fat ratio than men, and who pass on the chemical load to fetuses and babies through the placenta and breast milk (Physicians for Social Responsibility 2001). That this different experience registers biologically, however, does not equate with it being essentializing.

Another example of women's bodies being more vulnerable than men's can be seen from excess mortality related to heatwaves. Data for the French heatwave in 2003 show a 75% higher mortality rate for women than for men, which cannot be explained solely by women's greater longevity (Fouillet et al 2006). Likewise, in London in 1995, mortality was more noticeable among women (23.3% excess deaths compared with 8.3% excess deaths for men). There is no simple explanation for this, but Rooney et al (1998) suggest a combination of poverty, deprivation, and vulnerability to associated air pollution. Other reasons put forward included the vulnerability of living alone (Fouillet et al 2006) and the increased difficulty that women over 60 have in regulating their internal temperature (Kaciuba-Uscilko and Grucza 2001 in Laadi et al 2006). These data suggest that global warming will have a disproportionate effect on women, compared with men in any one community, as extreme weather episodes become more frequent, although this is not a consequence discussed in climate justice discourse.

The Domestic Scale

Social relations within the home (as a site for reproduction and recreation as well as, increasingly, paid work) are also neglected in many environmental justice analyses. Turner and Brownhill (2006:92) describe wives and children undertaking unwaged caring and domestic work as "shadow workers", reflecting the relegation of unpaid work in the private sphere. They argue that these workers have a distinct relationship with the environment as a result of their unequal position:

> ... the shadow worker reproduces not only labour power but also to some degree both the natural and built environments. Secondly, while the proletarian's wage entails a degree of power over both the commodity nexus and the shadow labour of his wife and kin, the shadow worker to a much larger degree depends on access to non-commodified or public goods both for subsistence and for expanding her autonomy from wage-earning kin (and capital). Thirdly ... the ideological construction of those identity groups assigned to shadow work is deeply linked to constructions of nature. (Turner and Brownhill 2006:92)

Turner and Brownhill further warn against a failure to acknowledge the full range of ways in which these shadow workers are marginalized, one result of which is a failure to fully appreciate the nature of environmental injustice. While the home-as-workplace is technically covered by health and safety workplace legislation, which requires monitoring by employers (HSE 2006), the home remains a largely unsupervised and potentially more hazardous space. It is here where much female labour (unregulated and unpaid as in housework and care of dependants; unregulated and low paid in the case of domestic workers) takes place.

Despite the proportion of women in paid work increasing, the majority of unpaid caring and daily household chores continue to be undertaken by women (United Nations Development Programme—UNDP 2007). Robeyns (2008) also points to the unequal distribution of work in the household, which, she argues, has a significant impact on the relative capabilities of household members. It is remarkable that data from North America and Europe reveal that cooking, cleaning, provisioning, food growing, waste disposal, caring for the sick, frail, young and otherwise dependent continues to be mostly done by women, despite the steadily increasing presence of women in the paid workforce (UNDP 2007). This brings women into direct contact with polluting activities, and necessitates a navigation between the needs and health of the cared for, economics, time, and environmental considerations. Although poor communities are more exposed to environmental problems relative to richer communities, within any given community it tends to be women who experience environmental problems disproportionate to their poverty alone.

Sze (2004) draws attention to the way in which disease is gendered in that caring for sick family members tends to fall to the mother in the household, and that judgment on the causes of disease sometimes brings into question the quality of this caring and mothering. Her analysis highlighted the intersectionality of race and gender in an environmental justice campaign against asthma-related pollutants. Asthma disproportionately affects the low-income minority—particularly African-American children in urban areas in the United States (as elsewhere), and has been rising sharply. Sze summarized scientific medical evidence which points to both indoor and outdoor pollution as being causes of asthma, but she particularly focuses on the way in which, while government health authorities and corporations claimed "no scientific proof" linking polluting facilities with asthma, the government "Healthy Homes" programme was designed to inspect "usually poor households with asthmatic children". Sze argues that the implication that "poor women or women of color have bad housekeeping practices" and that mothers are to blame for their children's asthma takes the responsibility away from corporations, inadequate government regulations or systemic failures (Sze 2004:188). Indeed, it is an irony that the women's campaign to reduce pollution at source resulted in the authorities reflecting back this concern as an (the mother's) inadequacy.

Daily housework and caring responsibilities are being augmented by demands that governments are placing on households to, for example, recycle, which, since these practices are mostly undertaken by women (Buckingham, Reeves and Batchelor 2005; Defra 2002; MacGregor 2006), adds to their "care burden". In the West, environmental considerations might include whether to buy organically certified food and biodegradable cleaning products, whether to compost organic waste and recycle other waste, whether to use reusable nappies, menstrual products and cleaning cloths or to opt for disposable alternatives. Environmentally related campaigns such as "slow food" and "zero waste" are laudable, but it must be recognized that until there is a fair and equitable division of labour in the home, the majority of the burden of this transition will fall to women. As MacGregor has pointed out "with the rise of public campaigns for environmental awareness, those who manage households . . . are expected to be more diligent in adopting time consuming green practices like recycling and precycling" (2006:69). This is a wider conception of "environmental injustice" than has hitherto been used, and in the spirit of "gender mainstreaming" it should be incumbent on government departments and environmental campaigning groups to calculate the gender impact of their campaigns— or, even better, to urge their members to "share the housework". For an example of how this could emerge, see Merchant's (2005) work on "partnership ethics". To date, however, there has been what MacGregor describes as a "failure of masculinist environmentalisms to address the gendering of experience and responsibility in the domestic sphere" (MacGregor 2006:61).

UK "nappy politics" illustrates the relationship between inequalities in the division of labour and environmental considerations, as well as the way in which the domestic scale is imbricated with the national. Although disposable nappies constitute approximately 3% of the UK waste stream and thereby pose an environmental problem, new mothers are targeted by manufactures who distribute a "bounty pack" through the British National Health Service (NHS), containing disposable nappies and toiletries using artificial chemicals (against the NHS's own advice that only water should be used on a baby's skin until she or he is 6 months old; Women's Environmental Network—WEN 2008). New mothers are a vulnerable and impressionable "market" in their anxiety to be "good" mothers, a point which manufacturers of disposable nappies exploit in their advertising and marketing campaigns. On the other hand, these women are urged by women's environmental and health campaigning groups to consider the links between their children's and their environment's wellbeing, and adopt the use of reusable nappies. This conflict was aggravated in 2005, when the Environment Agency released its life cycle analysis of reusable and disposable nappies, which alleged that disposable nappies were not more harmful to the environment than reusable nappies, influencing new parents' purchasing decisions and local authority policy. This life cycle analysis is generally thought to have been flawed (see, for example, Hickman 2005; WEN 2008) both on the assumptions made by the study,[5] and the sampling frame (2000 parents using disposable nappies, 117 parents using reusable nappies; Environment Agency 2005).

Like the slow food and zero waste movements identified earlier, the real nappy campaign, while empowering mothers to make well informed decisions (WEN 2008), increases the burden of household and caring work from both a physical and an emotional perspective. Consumer empowerment is double edged: women, who increasingly make household purchasing decisions (OECD 2008), become the battleground of emotionally laden consumer campaigns by both product manufacturers and environmental campaigners. In the absence of a fair division of domestic and caring labour, this produces a particular kind of environmental injustice in which environmental benefit is accrued at the expense of women's time and physical and emotional labour. Social injustice at the household scale (the division—or as Sen (2001) puts it—the accumulation of labour) needs to be factored into environmental justice considerations.

The relationality of scale, which emerges from this consideration of the body and the household, challenges the marginalization of "women's interests" as "inward" or the "politics of the personal" (Marston and Smith 2001; Salleh 1997:11). Brenner's questioning of the household as a valid scale—a "singular politics of scale" (2001:599)—denies household boundaries the porosity allowed to other scales. In contrast, Krall explains the porosity of the boundary of micro-scales, particularly visible in the female body, which "bears life" and food (1994:235). She also argues for better recognition of the boundary or "border spaces between the public and private realms" which enable those who inhabit these spaces (arguably more often women) to "provide a pragmatic, moral critique of both" (Krall 1994:236).

Gendering of Institutions and Organizations

While the distributional aspects of environmental justice focus on the unfair allocation of environmental problems and benefits, procedural aspects concern access to decision-making processes which govern these distributions. According to Young, "issues of the just organization of government institutions and just methods of political decision making rarely get raised" and inequalities are reinforced by decision-making structures which operate "to reproduce distributive inequality" (Young 1990:22–23). Gender inequalities in governing bodies have long been observed, and it has been noted that this has an impact on the nature of decisions made (Bhattar 2001; Buckingham et al 2005; Equality and Human Rights Commission 2008; Sealy and Singh 2008). This persistent inequality has led to the establishment of the Global Gender and Climate Alliance, set up jointly by the United Nations, the IUCN

(International Union for Conservation of Nature) and WEDO (Women's Environment and Development Organization), to try to raise the profile of gender issues in climate change policymaking (Aguilar 2009). Less attention, however, has been paid to professional environmental and environmental justice campaigning organizations, which often appear to demographically mirror the employment profiles of the businesses and institutions they challenge. Through Chopra and Duraippah's (2008:368) account of the recent institutionalization of organizations such as WWF, they question the ability of organizations able to institutionalize to break free of prevailing social norms and structures.

Gender mainstreaming[6] has been an attempt to draw more women into decision making, though its impact has been limited. For example, the Environment Agency, referred to above, is not an organization that has embraced gender mainstreaming—a practice statutorily required by European Law (CEC 1996)—as an interview with their equal opportunities officer pointed out.

Gender mainstreaming has been incorporated into international environmental legislation since 1995 (Morrow 2006) to ensure that institutional decisions do not have a disproportionately negative effect on one gender (usually taken to mean women and girls), and to draw more women into representative and effective decision making (see European Commission 1999; George 2007; Morrow 2006; United Nations 2005; World Bank 2002; and below in this chapter). While, technically, environmental decision making in the European Union is subject to gender mainstreaming, these policies are insufficient to ensure gender justice with regard to the environment, or decision making more generally. This is emphasized by research into gender mainstreaming European municipal waste management in the EU (Buckingham, Reeves and Batchelor 2005), which found that the ways in which municipal waste management was organized and gendered made a difference to waste strategies adopted, and, ultimately, the amount of waste disposed of. Four municipal waste authorities in the UK, Ireland and Portugal were examined in depth to find out how gender was being incorporated in their waste management strategies, and the scope for doing so. The importance of institutional actors was clearly identified with authorities employing women from diverse backgrounds (including education and marketing) to lead on recycling having a much stronger recycling performance than authorities in which waste management remained under the control of older men from an engineering background. Their preferred strategies were high-technology developments such as incineration, themselves the source of local protests—often dominated by women—against environmental injustice. Municipalities performing more strongly on recycling were also more likely to have better and more structured representation of women through public participation (Buckingham, Reeves and Batchelor 2005). Elsewhere, evidence suggests that while men favour technology in environmental problem solving, women tend to prioritize "soft" strategies such as changing lifestyles and consumption patterns which, conventionally, are given less prominence and significantly less financing (Johnsson-Latham 2007). In some ways, this is an unresolved contradiction, given the disproportionate impact these strategies have on women as the main, unpaid, domestic and care workers, as discussed earlier.

Conversely, Johnsson-Latham noted positive impacts where institutions have linked gender with their environmental work, particularly through initiatives set up by women, such as the Network of Women Ministers of the Environment (NWME) established in 2002 to promote women's participation in environmental policies and to enhance a gender perspective in national and international environmental decision making. Given that environmental justice campaigning seeks to influence policy, and becomes increasingly adopted by formal structures of government (see, for example, Greater London Authority 2004), the gendered nature of policymaking is important as it is likely to have a bearing on decisions and action taken.

Environmental Non-Governmental Organizations

Since environmental non-governmental organizations (hereafter ENGOs) are increasingly drawn into policymaking circles through a process Carter (2001) describes as "institutionalization", it is appropriate to examine the degree to which they reflect, in staffing and organizational terms, the organizations they are working with (see also Chopra and Duraippah 2008). The ENGOs being called on as a resource for decision making at national and international scales are generally those which are best able to dedicate time, staff and facilities to these activities. This contribution to governance is partial and inequitably drawn from the ENGO sector, which has implications for campaigns and strategies developed, and decisions taken. According to Salleh (2008), most NGO campaigning and policy formulation is conducted within masculinist neoliberal terms of reference.

Even if smaller, women-focused organizations were called as expert witnesses (and one of the author's experience of trusteeship with a leading national women's environmental organization suggests that they are frequently overlooked), they are unlikely to have the resources to enable them to participate in activities for which there is no financial recompense. The Women's Resource Centre (2006) has recently drawn attention to data which demonstrate that charities specifically working on behalf of women are disadvantaged relative to other charities, with women's organizations comprising 7% of all voluntary organizations in the UK but receiving only 1.2% of government funding to charities, and even less from independent funders. Since environmental charities receive relatively small charitable donations, with New Philanthropy Capital (2007) reporting that the hundred largest charitable trusts donate just 2% to environmental causes, this bodes particularly ill for women's environmental campaigning. While the organizations discussed below are not exclusively focused on environmental justice, it is increasingly the case in the UK that environmental NGOs use environmental justice as a campaigning hook, and, by virtue of their consultative status, they are powerful determinants of whether (and which) environmental justice issues are taken up in popular policy discourse.

Of the eight representative European families of sectoral NGO groups that formed the EU Civil Contact Group in 2008, a prerequisite of which is to be committed to equal opportunities, only two were headed by women (one was the European Women's Lobby, which was the only group to have a majority of women members). All of the chief executive officers of organizations comprising the environmental sectoral group advising the European Commission—the "Green Ten" were men (see Table 1). This does not necessarily imply that these groups do not take women, or gender, into account (gender does, in fact, feature on Friends of the Earth Europe's website as a campaigning issue under environmental justice). However, as with analyses of women's representation on government bodies, the boards of multi-national

Table 1: Gendering of ENGOs representing the Green Ten, advising the EU

Organization	Chair of governing board	% Board male	CEO
European Environmental Bureau	Male (president)	57	Male
Birdlife International European Community Office	Male	74	Male
Climate Action Network Europe	Male	63.5	Male
European Federation for Transport and Environment	Male	63.5	Male
Friends of the Earth Europe	Male	60	Male
Friends of Nature	Male	75	Male
Greenpeace European Unit (International)	Female (international)	66	Male
WWF European Policy Office	Male	61	Male
Health and Environment Alliance	Female	29	Male
CEE Bankwatch Network	–	–	Male

Sources: Bankwatch (2008), Birdlife International (2008), Climate Action Europe (2008), European Environmental Bureau (2008), European Federation for Transport and Environment (2008), Friends of Nature (2008), Friends of the Earth, Europe (2006, 2008), Greenpeace International (2008a, 2008b), Health and Environment Alliance (2008), WWF (2008a, 2008b).

Table 2: Gendering of largest British environmental NGOs

Organization (number of employees)	Chair of governing board	% Board male	Chief executive officer
Royal Society for the Protection of Birds (1300)	Male	88	Male
WWF UK (295)	Male	78	Male
Carbon Trust (151)	Male	70	Male
Greenpeace UK (117)	Female	42	Male
Friends of the Earth England (99)	Male	62	Male
National Trust (84)	Male	82	Female

Source: Kulcur (2008) (and annual reports 2008)

companies and other major employers with significant impact on people's everyday lives, it suggests underlying influences which favour the hiring of men to senior positions, and which may well have an impact on strategies, campaigns and advice given (Equality and Human Rights Commission 2008). A similar profile can be noted in the largest British ENGOs (see Table 2; Kulcur 2008), only two of which have women chairs, and one a woman chief executive officer. With the exception of Greenpeace, the boards are heavily dominated by men (from 63% to 88%).

This understanding of a balance of power in favour of men is reinforced by a survey undertaken by the UK Directory of Social Change, which asked the question: "Do women control the voluntary sector?" Despite 68% of the voluntary sector workforce being women (which incidentally throws the male domination of senior posts in ENGOs into even sharper relief), and 54% volunteers, "overwhelmingly the respondents felt that key decisions remain the province of males" (Directory of Social Change 2006).

Fincher draws attention to the fact that the state is shaped by ways in which it is contested by activitist organizations; how these activist organizations are spatialized and gendered is important to this. Organizations outside the state "help form the actions of the state, shift its practices and alter its interpretations" by "interacting with actors in state bureaucracies in diverse and interesting ways" (Mountz 2003:18 in Fincher 2007). Fincher argues that it might be useful to "pay more attention to the ways institutions are shaped by their members' interactions with groups and individuals from 'outside' their bureaucratic borders" (2006:13). This is increasingly salient in the environmental sphere, where, as discussed above, campaigning groups, which are demonstrably dominated by men in senior management and governance positions, are increasingly being incorporated by governments to develop environmental strategies. Analysis of a key resource for campaigning organizations, the Worldwatch Institute, has revealed its signal failure to consider gender as an issue in consumption issues (Johnsson-Latham 2007). Similarly, Lina Sommestad, Minister for the Environment in Sweden between 2002 and 2006, reported a male colleague's response to an initiative on women and the environment as hostile: "What is this? I cannot see why environment has anything at all to do with women" (Sommestad 2007). This attitude has also been exposed in interviews with ENGOs in the UK, in which senior managers failed to see the relevance of gender-sensitive campaigns (Kulcur 2008), and in the European waste management research cited earlier, with ENGOs demonstrating a singular lack of awareness of what a gender analysis could offer. When the director of one of the leading environmental organizations for Portugal was asked about women's involvement in waste campaigns, his response was:

taking in consideration our lack of time and resources, and the priorities that are in that line, I sincerely think (and this is a personal opinion . . .) that [including women] is not a priority issue regarding waste. (Buckingham et al 2004:appendix)

NGOs can "become sites reinforcing the status quo because of their inaction", and consequently cannot be assumed to be "progressive space for social, political and economic activities" (Ruwanpura 2007:318). She found in her work in Sri Lanka that NGOs often worked within the cultural constraints of bureaucracies and institutions, thereby ignoring gender inequalities/atrocities, construed as "culturally acceptable (Ruwanpura 2007:325). The evidence presented above suggests that her argument, that neglecting an analysis of how NGOs integrate (or otherwise) political strategy, power and feminism into an analysis of civil society and how this affects the "structural exclusion of women" (Ruwanpura 2007:327), is relevant for environmental NGOs working in Europe and the USA. Examining the gendered attitudes of mainstream ENGOS to environmental justice is, we would argue, critically important, given their influence in public discourse. For example, Shrader-Frechette (2006) argues that more mainstream groups in the USA, such as the Sierra Club, have failed to support environmental justice campaigns because their own structures mirror those in society at large. Such a failure is influential in whether and how environmental justice is taken up by government and other institutions.

Brulle and Essoka (2005) cite Rios's research into the environmental justice movement itself which she found to be more like conventional social movements, with institutional structures and NGO funding. This distanced decision making from its grassroots membership which, as earlier citations have already demonstrated, are often dominated by women, and especially women of colour (Di Chiro 1998; Sze 2004; Taylor 1997). Brulle's own research found environmental justice organizations to have lower levels of democracy than some mainstream environmental groups, whose governance structures they shared (Brulle and Essoka 2005:211–215). Brulle and Essoka (2005:217) call for a "broad based effort" to "ensure the organizations of this movement are authentic representations of the communities in which they are based". Agyeman's work with environmental justice campaigning groups in Boston discusses the difference in staffing, with the community organizers reflecting the ethnic diversity of the neighbourhoods in which they are working, while the office staff are more likely to be white and college educated (Agyeman 2005:142). In his analysis gender balance has less consideration, with men more likely to dominate in the office. A quote from a campaign group member concerning the perceived "inauthenticity" of a black woman who was a "Berkeley trained lawyer" and "dating white guys" (Agyeman 2005:142) says as much about gender relations in the organization as it does about race. These analyses of power in organizations campaigning against environmental injustice are particularly important in view of the dominance of grassroots activity originating with women (Schrader-Frechette 2002; Taylor, 2000). As Cheng (1996:viii) suggests, a possible transformation can only be assessed if the organization as a whole is examined.

Arguably civil society organizations such as the Women's Environment and Development Organization (WEDO), Women in Europe for a Common Future, ENERGIA, LIFE, the Gender and Water Alliance, Women Organizing for Change in Agriculture and Natural Resource Management (WOCAN), and WEN all play important, but in light of the foregoing discussions, limited, roles in bringing a gender perspective to the political arena, which can have an impact on environmental justice.

Conclusions: Gender Sensitive Ways Forward for Environmental Justice

Both the environmental justice movement and ecofeminism have drawn attention to the way in which the traditional environmental movement has tended to assume that environmental damage impacts on people universally. While environmental justice campaigning has clearly revealed this as a fallacy, organizations lobbying for environmental justice generally fail to investigate how environmental racism might itself be fractured by gender and other variables, even though much of the grassroots campaigning in the USA is undertaken by women, often primarily concerned with environmental health issues (Sze 2004). Indeed, women fighting for environmental justice in the USA in African-American and other disadvantaged communities have been vilified on the basis of their sex (see Gibbs 1996; Kurtz 2007), and their gender roles (Sze 2004). The sexist language used in challenges to women's claims for environmental justice can also become personalized. Kurtz, for example, quotes ways in which state officials publicly denigrated the protest group she was researching as "a bunch of gray-haired ladies" and "hysterical housewives" thereby "construct[ing] women activists not only as external to the public sphere, but also as people unconstrained by the rationality of public sphere identification and behaviour..." (Kurtz 2007:417). Kurtz draws parallels with the criticisms leveled at Lois Gibbs when she organized the fight against Love Canal in the 1970s, which perjoratively accused her of presenting "housewives' data".

As with many political movements fought in the name of "liberation" (see Mehdid on Algeria, 1996; Lievesley on Cuba, 2006), within the environmental movement gender difference has been suppressed in the name of "greater humanity, community or class" (Salleh 1997: 6), or as the quote from the Portuguese ENGO suggested earlier, by "lack of time". Ruwanpura challenges the faith that "many, including feminists" have placed "in these [NGO] actors for creating a progressive space ...", arguing that the NGOs "do not necessarily critically examine the neo-liberal agenda for its lack of attention to the nationalist and gender dynamics of local politics". This, she suggests, runs the risk of reinforcing "prevailing ... gender dynamics" in that NGOs ignore the ways in which women from each community are marginalized. Indeed, in her conclusions, Ruwanpura suggests that to neglect the intersections of political strategy, power and feminism, "would be to inadvertently promote everyday forms of structural exclusion of women ..." (2007:317). In an indirect appeal to intersectionality, Salleh invites the environmental justice movement to ask "where does the impact of class or race end, and gender effect begin" (1997:7). Although speaking with regard to Native people, Krall's assertion that "Even actions performed for the common good cannot be justified if they deny some the right to be", draws attention to the danger of the best motivated of campaigning activity ignoring the complexity of social marginalization (Krall 1994:167).

This chapter has observed a paucity of gender analysis in the literature on environmental (in)justice, and a failure of many campaigns against environmental injustice to engage with their inherent gender injustices. From the evidence we have considered it appears that gender considerations—and women's involvement in environmental justice, as often as not expressed as health campaigns—diminish, the further removed these campaigns are from their grassroots. The scales at which these injustices are manifested are also important considerations. In terms of both impacts and campaigns, the body and the household are key, but often marginalized, sites of environmental injustice, while in terms of campaigns, there is a distinct attenuation of women's presence the further from the grassroots one travels. The lack of women in organizations that arguably have the power to challenge environmental injustice at the national scale reinforces gender inequalities and limits the effectiveness of environmental justice. We argue that the "first environment" of the body (Sze 2006) needs to be articulated much more explicitly as a site of environmental injustice and that the domestic division of labour constitutes a social injustice which is compounded by environmental considerations. The persistently unequal position of women, even in the technologically advantaged countries of North America and Europe, is both highlighted and reinforced by environmental problems. Unless the gender dimension of these problems is addressed in strategies put forward to overcome them, the social injustice of gender inequality, compounded with inequalities of race, ethnicity, poverty and age, will endure. We suggest that social relations within the household and in the workplace, as well as gender imbalances in influential posts in institutions and organizations which shape environmental justice discourse, are likely, based on an analysis of gender and decision making, to have an impact on the decisions made (Equality and Human Rights Commission 2008; Barsh, Cranston and Craske 2008). This imbalance, we further suggest, is compounded by an environmental sector whose staffing profiles closely mimic those in the governments and industries they lobby and campaign against. The impacts of such staffing patterns are under researched in environmental campaigning and decision making, and we suggest that this needs researching more closely (see Kulcur 2008).

For these reasons, we argue that a focus on either race/ethnicity or poverty, as delineations of environmental injustice, will not be sufficient to ensure that gendered injustices as a result of environmental problems are resolved. While there is vibrant women's involvement in grassroots environmental justice and health campaigning (Agyeman 2005; Shrader-Frechette 2002; Sze 2004; Taylor 1997, 2000), this chapter has also revealed underfunding in the charitable sector exclusively focused on women (relative to nongender specific groups, or groups focused on animals, wildlife, buildings and so on). The examples we have drawn on, based on our fieldwork, research projects and environmental activism, suggest that the ways in which organizations are structured affect how they deal with both gender difference and environmental injustice. We would also argue that environmental justice can only be achieved if environment and gender as well as other structures of injustice are identified and recognized at all scales. Oppression at all scales—from the body and household, through to the national and international—needs better articulation and understanding within the organizations campaigning for environmental justice. For this, further analysis of the links between environmental injustices and gender oppression, and a commitment to gender equality, are urgently needed.

Acknowledgements

With thanks to the Department of Resource Management and Geography at the University of Melbourne which hosted a period of research leave for Susan Buckingham during which this chapter was first written, and to Ariel Salleh who generously offered her time and critical comments on an early draft. Thanks also to the anonymous reviewers whose trenchant comments were valuable in strengthening the arguments of the chapter.

Endnotes

[1] The concept of race is used in US environmental justice campaigning, while in the UK, ethnicity is used.

[2] While there is a substantial literature on global environmental injustices, and many environmental justice campaigns highlighting North/South injustices, this chapter focuses on the environmental justice movement in the USA and Europe, and draws specifically on research conducted in Europe (primarily the UK).

[3] Robeyns uses Amartya Sen's "capabilities approach", which focuses on the social, environmental and personal factors that enable people to convert goods and services into useful outcomes for themselves. It is predicated on an assumption of heterogeneity whereby each individual has a unique bundle of capabilities with which to effect this conversion. Importantly, this presupposes that it is capabilities such as education, health, social networks, respect from others in society and so on which determine how effectively goods and services can be used. While Sen's analysis primarily concerns people in the Global South, Robeyns applies this concept in "affluent and technologically advanced economies" (2008:83) to explore its specific utility in addressing feminist concerns.

[4] REACH is the single integrated system for the 'registration, evaluation and authorization of chemicals' which the European Union has established.

[5] For example, the research underpinning the research presupposed that parents using cloth nappies would be using a washing machine dating from 1997; that families would use an average of 47 nappies per child (Guardian columnist, Leo Hickman suggests that this is more than double the number used in his family, (Hickman, 2007)); and that these nappies would only be used for one child, would be ironed (as Hickman wrote, incredulously 'who on earth irons their nappies?'), tumble dried and washed at 90 degrees C (even though, as WEN points out, the reusable nappy manufacturers recommend 60C). NB The Environment Agency published a revised analysis in October 2008 which is more equivocal concerning the comparable benefits of reusable and disposable nappies.

[6] Karen Morrow suggests that gender mainstreaming (GM) takes two forms: institutional and instrumental. In the former, GM is "recognized as an integral part of the mainstream institutional agenda" and it is used to make institutional change; in the latter, "gender-informed approaches are brought to bear on subject specific laws and policies and upon their implementation" (2006:39).

References

Aguilar L (2009) Women and climate change: Vulnerabilities and adaptive capacities. In *2009 State of the World. Into a Warming* (pp 59–62). Washington: Worldwatch Institute

Agyeman J (2000) *Environmental Justice. From the Margins to the Mainstream*. London: Town and Country Planning Association

Agyeman J (2005) *Sustainable Communities and the Challenge of Environmental Justice*. New York: New York University Press

Agyeman J, Bullard R and Evans B (eds) (2003) *Just Sustainabilities: Development in an Unequal World*. London: Earthscan

Bankwatch (2008) Bankwatch overview. http://www.bankwatch.org/about/ Accessed 23 February 2008

Barsh J, Cranston S and Craske R (2008) Centered leadership: How talented women thrive. *The McKinsey Quarterly* No 4

Bhattar G (2001) Of geese and ganders: Mainstreaming gender in the context of sustainable human development. *Journal of Gender Studies* 10(1):17–32

Birdlife International (2008) Staff. http://www.birdlife.org/eu/eco_staff.html Accessed 23 February 2008

Bradshaw J, Finch N, Kemp P A, Mayhew E and Williams J (2003) *Gender and Poverty in Britain*. Manchester: Equal Opportunities Commission

Brenner N (2001) The limits to scale? Methodological reflections on scalar structuration. *Progress in Human Geography* 25:591–614

Brulle R J and Essoka J (2005) Whose environmental justice? An analysis of the governance structure of environmental justice organisations in the United States. In N D Pellow and R J Brulle (eds) *Power, Justice and the Environment, a Critical Appraisal of the Environmental Justice Movement* (pp 205–218). London: MIT Press

Buckingham S (2006) Environmental action as a space for developing women's citizenship. In S Buckingham and G Lievesley (eds) *In the Hands of Women: Paradigms of citizenship* (pp 62–94). Manchester: Manchester University Press

Buckingham S (2007) Microgeographies and microruptures. The politics of gender in the theory and practice of sustainability. In R Krueger and D Gibbs (eds) *The Sustainable Development Paradox; Urban Political Economy in the United States and Europe* (pp 66–94). New York: Guilford

Buckingham S, Reeves D and Batchelor A (2005) Wasting women: The environmental justice of including women in municipal waste management. *Local Environment* 10(4):427–444

Buckingham S, Reeves D, the Women's Environmental Network, Batchelor A and Colucas S (2004) *Research into Gender Differentiated Impacts of Municipal Waste Management Planning in the European Union, Final Report to the CEC, DG–Environment*. London: Brunel University

Bullard R D (1999) Dismantling environmental racism in the USA. *Local Environment* 4(1):5–20

Carbon Trust (2009) The board. http://www.carbontrust.co.uk/about/people Accessed 9 July 2009

Carter N (2001) *The Politics of the Environment: Ideas, Activism, Policy*. Cambridge: Cambridge University Press

CEC (Commission of the European Community) (1996) *Gender Mainstreaming*. Brussels: CEC

Checker M (2005) *Polluted Promises. Environmental Racism and the Search for Justice in a Southern Town*. New York: New York University Press

Cheng C (1996) *Masculinities in Organizations, Research on Men and Masculinities*. London: Sage

Chopra K and Duraippah A K (2008) Operationalizing capabilities in a segmented society: The role of institutions. In F Comim, M Qizilbash and S Alkire (eds) *The Capability Approach, Concepts, Measures and Applications* (pp 362–382). Cambridge: Cambridge University Press

Climate Action Europe (2008) Climate Action Europe: Team and contacts. http://www.climnet.org/about/team.htm Accessed 23 February 2008

Cuomo C (1994) Ecofeminism, deep ecology and human population. In K J Warren (ed) *Ecological Feminism* (pp 88–105). London: Routledge

Defra (2002) *Survey of Public Attitudes to Quality of Life and to the Environment–2001*. London: Defra

Di Chiro G (1998) Environmental justice from the grassroots: Reflections on history, gender and expertise. In D R Faber (ed) *The Struggle for Environmental Democracy: Environmental Justice Movements in the United States* (pp 104–136). New York: Guilford Press

Directory of Social Change (2006) Majority say women do not control the voluntary sector. http://www.dsc.org.uk/charityexchange/news_archives/survey_results1006.html Accessed 21 February 2008

Dunnion K and Scandrett E (2003) The campaign for environmental justice in Scotland as a response to poverty in a northern nation. In J Agyeman, R Bullard and B Evans (eds) *Just Sustainabilities: Development in an Unequall World* (pp 311–322). London: Earthscan

EEA (European Environment Agency) (2003) *Environmental Assessment Report*. Copenhagen: EEA

Environment Agency (2005) *Life Cycle Assessment of Disposable and Reusable Nappies in the UK*. Bristol: Environment Agency

Equality and Human Rights Commission (2008) *Sex and Power 2008*. London: EHRC

European Commission (1999) *Gender Mainstreaming 2/1999*. Brussels: CEC

European Environmental Bureau (2008) Board. http://www.eeb.org/board/Index.html Accessed 23 February 2008

European Federation for Transport and Environment (2008) Staff and board. http://www.transportenvironment.org/module-htmlpages-display-pid-3.html Accessed 23 February 2008

Fincher R (2006) Space, gender and institutions in processes creating difference. *Gender, Place and Culture* 14(1):5–27

Fouillet A, Rey G, Laurent F, Pavillon G, Bellec S, Guihenneuc-Jouvaux C, Clavel J, Jougla E and Hemon D (2006) Excess mortality related to the August 2003 heat wave in France. *International Archives of Occupational and Environmental Health* 80(1):16–24

Friends of Nature (2008) Board. http://www.nfi.at/index.php?option=com_content&task=view&id=5&Itemid=11, Accessed 4 November 2008

Friends of the Earth Europe (2008a) Contacts. http://www.foeeurope.org/about/contact.htm, Accessed on 23 February 2008

Friends of the Earth Europe (2008b) Annual review. http://www.foeeurope.org/publications/2007/FoEE_annual_review2006.pdf Accessed 20 February 2008

Friends of the Earth Scotland (2000) *The Campaign for Environmental Justice*. Edinburgh: Friends of the Earth Scotland

George G R (2007) Interpreting gender mainstreaming by NGOs in India: A comparative ethnographic approach. *Gender, Place and Culture* 14(6):679–701

Gibbs L (1996) *Love Canal: the Story Continues*. Gabriola Island, British Columbia: New Society Press

Godfrey P (2006) Essentialism and the semantics of resistance: A rejoinder to Jesse Goldstein. *Capital, Nature, Socialism.* 17(4):103–104

Goldstein J (2006) Ecofeminism in theory and praxis. *Capital, Nature, Socialism* 17(4):96–192

Greater London Authority (2004) *Environmental Justice Report*. London: GLA

Greenpeace International (2008a) Executive director. http://www.greenpeace.org/international/about/how-is-greenpeace-structured/management/executive-director Accessed 16 April 2009

Greenpeace International (2008b) International board. http://www.greenpeace.org/international/about/how-is-greenpeace-structured/governance-structure/board, Accessed 16 April 2009

Health and Environment Alliance (2008) Executive committee. http://www.env-health.org/r/138, Accessed on 23 February 2008

Hickman L (2005) It won't wash. *The Guardian* 20 May

Johnsson-Latham G (2007) *A Study on Gender Equality as a Prerequisite for Sustainable Development*. Stockholm: Environmental Advisory Council, Sweden

Kaciuba-Uscilko H (2001) Gender differences in thermoregulation. *Current Opinions in Clinical Nutrition and Metabolic Care* 4:533–536

Krall F R (1994) *Ecotone: Wayfaring on the Margins*. Albany, NY: SUNY Press

Kulcur R (2008) Gender and environmental movement. Paper presented at RGS-IBG Conference Session Gender Matters: Postgraduate Reflections. London, UK, August

Kurtz H E (2007) Gender and environmental justice in Louisiana: Blurring the boundaries of public and private spheres. *Gender, Place and Culture* 14(4):409–426

Lee J-A (2007) Gender, ethnicity, and hybrid forms of community-based urban activism in Vancouver, 1957–1978: The Strathcona story revisited. *Gender, Place and Culture* 14(4):381–407

Lievesley G (2006) Identity, gender and citizenship: Women in Latin and Central America and Cuba. In S Buckingham and G Lievesley (eds) *In the Hands of Women: Paradigms of Citizenship* (pp 127–162). Manchester: Manchester University Press

London Sustainability Exchange (2004) *Environmental Justice in London. Linking the Equalities and Environmental Policy Agendas*. London: London Sustainability Exchange

MacGregor S (2006) *Beyond Mothering Earth. Ecological Citizenship and the Politics of Care*. Vancouver: University of British Colombia Press

Marston S (2000) The social construction of scale. *Progress in Human Geography* 24(2):219–242

Marston S and Smith N (2001) States, scales and households: limits to scale thinking? A response to Brenner. *Progress in Human Geography* 25:615–619

McDowell L (1999) *Gender, Identity and Place. Understanding Feminist Geographies*. Cambridge: Polity

McDowell L (2004) Work, workfare, work/life balance and an ethic of care. *Progress in Human Geography* 28(2):145–163

Mehdid M (1996) Engendering the nation state: Women, patriarchy and politics in Algeria. In S M Rai and G Lievesley *Women and the State: International Perspectives* (pp 78–102). London: Taylor and Francis

Merchant C (1996) *Earthcare: Women and the Environment*. London: Routledge

Merchant C (2005, 2e) *Radical Ecology*. London: Routledge

Morrow K (2006) Gender, international law and the emergence of environmental citizenship. In S Buckingham and G Lievesley (eds) *In the Hands of Women: Paradigms of Citizenship* (pp 33–61). Manchester: Manchester University Press

National Trust (2009) *Annual Report and Accounts*. http://www.nationaltrust.org.uk/main/w-trust Accessed 9 July 2009

New Philanthropy Capital (2007) *Green Philanthropy. Finding Charity Solutions to Environmental Problems, a Guide for Donors and Funders*. London: NPC

Nightingale A (2006) The nature of gender: Work, gender and environment. *Environment and Planning D: Society and Space* 24:165–185

Pellow N D and Brulle R J (2005) *Power, Justice, and the Environment. A Critical Appraisal of the Environmental Justice Movement*. London and Cambridge, MA: MIT Press

Physicians for Social Responsibility (2001) *Environmental Endocrine Disruptors: What Physicians Should Know*. http://www.psr.org/site/DocServer/Environmental_ Endocrine_Disruptors.pdf?docID=575 Accessed 21 February 2008

Plumwood V (1993) *Feminism and the Mastery of Nature*. London: Routledge

Pulido L (1996) *Environmentalism and Economic Justice. Two Chicano Struggles in the Southwest*. Tucson: University of Arizona Press

Robeyns I (2008) Sen's capability approach and feminist concerns. In F Comim, M Qizilbash and S Alkire (eds) *The Capability Approach, Concepts, Measures and Applications* (pp 82–104). Cambridge: Cambridge University Press

Rooney C, McMichael A J, Kmovats R S and Coleman M P (1998) Excess mortality in England and Wales, and in Greater London, during the 1995 heatwave. *Journal of Epidemiology and Community Health* 52(8):482–486

RSPB (2008) *The RSPB Trustees' Report and Accounts for the Year Ended 31 March 2008*. http://www.rspb.org.uk/images/Trusteesannualreportandaccounts-2007-08 Accessed 9 July 2009

Ruwanpura K N (2007) "Awareness and action": The ethno-gender dynamics of Sri Lankan NGOs. *Gender, Place and Culture* 14(3):317–333

Salleh A (1997) *Ecofeminism as Politics: Nature, Marx and the Postmodern*. London: Zed Books

Salleh A (2006) Embodying the deepest contradiction: A rejoinder to Alan Rudy. *Capital, Nature, Socialism* 17(4):115–124

Salleh A (2008) *Personal communication*

Schlosberg D (2007) *Defining Environmental Justice. Theories, Movements and Nature*. Oxford: Oxford University Press

Schwarte C and Adebowale M (2007) *Environmental Justice and Race Equality in the European Union*. London: Capacity Global

Seager J (1993) *Earth Follies: Coming to Terms With the Global Environmental Crisis*. London: Routledge

Sealy R and Singh V (2008) The importance of role models in the development of leaders' professional identity. In K James and J Collins (eds) *Leadership Learning: Knowledge into Action*. Basingstoke: Palgrave Macmillan

Sen A (2001) Many faces of gender inequality. Inauguration lecture presented to the Radcliffe Institute at Harvard University, April

Shakespeare T and Watson N (2002) The Social Model of Disability: An outdated ideology? *Research in Social Science and Disability* 2:9–28

Shrader-Frechette K (2002) *Environmental Justice, Creating Equality, Reclaiming Democracy*. Oxford: Oxford University Press

Sommestad L (2007) Women's leadership for the environment as a human right in Europe, Eastern Europe and beyond. Keynote paper presented at WAVE conference in Belgrade, Serbia, October. http://www.wecf.eu/cms/download/ 2007/Kenote_Address_Sommestad-WAVE_2007.pdf Accessed 6 June 2009

Sze J (2004) Gender, asthma politics, and urban environmental justice activism. In R Stein (ed) *New Perspectives on Environmental Justice. Gender, Sexuality, and Activism* (pp 177–190). New Brunswick, NJ: Rutgers University Press

Sze J (2006) Boundaries and border wars: DES, technology, and environmental justice. *American Quarterly* 58(3):791–814

Taylor D (1997) Women of color, environmental justice and ecofeminism. In K Warren (ed) *Ecofeminism: Women, Culture and Nature* (pp 38–81). Bloomington: Indiana University Press

Taylor D (2000) The rise of the environmental justice paradigm: Injustice framing and the social construction of environmental discourses. *American Behavioural Scientist* 43(4):508–580

Turner T E and Brownhill L (2006) Ecofeminism as gendered, ethnicized class struggle: A rejoinder to Stuart Rosewarne. *Capitalism Nature Socialism* 17(4):87–95

United Nations (2005) *Taking Action: Achieving Gender Equality and Empowering Women.* UN Millennium Project Task Force on Education and Gender Equality. London: Earthscan

United Nations Development Programme (2007) *Fighting Climate Change. Human Solidarity in a Changing World.* New York: UNDP

Walker G and Bulkeley H (2006) Geographies of environmental justice. *Geoforum* 37(5):655–659

Walker G, Mitchell G, Fairburn J and Smith G (2005) Industrial pollution and social deprivation: Evidence and complexity in evaluating and responding to environmental inequality. *Local Environment* 10(4):361–378

Warren K (ed) (1997) *Ecofeminism: Women, Culture, Nature.* Bloomington: Indiana University Press

Women's Environmental Network (2008) Environment Agency report reveals cloth nappies can save 40% carbon emissions. http://www.wen.org.uk/general_pages/Newsitems/pr_nappyreport17.10.08.htm Accessed 4 November 2008

Womens' Resource Centre (2006) *Why Women? The Women's Voluntary and Community Sector: Changing Lives, Changing Communities, Changing Society.* London: Women's Resource Centre

World Bank (2002) Gender mainstreaming strategy paper. http://www.worldbank.org/gender/overview/ssp/home/htm Accessed 23 October 2006

WWF (2008) About WWF. http://www.worldwildlife.org/about/boardlist.cfm Accessed 23 February 2008

WWF (2008) Board of trustees. www.panda.org/about_wwf/where_we_work/europe/what_we_do/epo/about_us/about_people/index.cfm Accessed 16 April 2009

Young I M (1990) *Justice and the Politics of Difference.* Princeton: Princeton University Press

Chapter 4

Acknowledging the Racial State: An Agenda for Environmental Justice Research

Hilda E. Kurtz

Introduction

The purpose of this chapter is to argue that critical race theory offers environmental justice (EJ)[1] scholars important insights into role of the state in both fostering and responding to conditions of racialized environmental injustice. I use a schematic review of the EJ literature in geography[2] to argue that EJ scholars implicitly acknowledge that racism is threaded through state practices, but do not consider the implications of a racialized state apparatus for EJ activism with any theoretical vigor. Without a more sophisticated consideration of the racialized nature of the state, our understandings of racially oriented EJ activism—its successes, failures and remaining possibilities with regard to the state—will remain woefully incomplete. Critical race theory offers insights into both why and how the state manages racial categories in such a way as to produce environmental injustice, and how the state responds to the claims of the EJ movement.

I draw on David Goldberg's (2002) argument in *The Racial State* and Omi and Winant's (1994) theory of *Racial Formation in the United States* to argue for a re-invigorated approach to EJ scholarship. As a critical race theorist and state theorist, Goldberg (2002) traces the history of the racial state, and demonstrates forcefully that the state plays a centrally important, if changing, role in creating and policing racial categories over time. Omi and Winant (1994) argue that the racial state and racial movements such as the EJ movement are linked in an historical framework of racial formation. Their analysis of the American civil rights movement highlights the importance of examining a social movement *in relation to* the state apparatus it attempts to change. Together, these works highlight the importance of race management to the conduct of the modern liberal state, however complicated and contradictory that might sometimes be. I draw on their insights to argue that EJ scholarship would benefit from a more robust understanding of the state *in relation to* environmental injustice and the social movement which seeks to redress it.

I do not wish to suggest that only people of color experience environmental injustice, as there is considerable evidence to the contrary. Efforts to cast the question of "justice to whom" in broadly inclusive terms take into account race, class, gender, age and future generations (Walker and Bulkeley 2006). I do, however, wish to move discussion of the role of the state in environmental injustice to consider not only the efficacy but also the genealogy of state initiatives such as Executive Order 12898, the National Environmental Justice Advisory Council (NEJAC), and the Interagency Working Group. As Omi and Winant's (1994) work underscores, such initiatives emerge from a pattern of conflict and accommodation between racial state apparatuses and the EJ movement. Closer attention to the interplay between the racial state and the EJ movement as a racial social movement will yield important insights into the conditions, processes, institutions and state apparatuses that foster environmental injustice and that delimit the possibilities for achieving EJ in some form or another.

In the next section I highlight three key tensions shaping the EJ literature and argue that in each of these arenas of debate, EJ scholars tend to overlook the significance of the state's role in shaping understandings of race and racism. I then draw on Goldberg's discussion of the racial state to suggest that we might fruitfully move beyond these three tensions. Finally, I draw from Omi and Winant's (1994) theory of racial formation in order to suggest ways to figure the racial state into future EJ research.

Three Key Debates in the Environmental Justice Literature

Geographic scholarship on EJ is shaped by several key tensions. A fundamental debate concerns whether racism (however construed) or capitalism is the primary cause of environmental injustice, and therefore the most appropriate target of EJ activism and critique. The racism/capitalism debate intersects with two other tensions that concern ways of knowing the problem of environmental injustice at all. First, expert knowledge(s) grounded in Census and other spatial data are in tension with lay and experiential knowledge(s) of the disparate impacts of pollution on peoples' health and livelihoods; fundamentally different ways of knowing shape often divergent perspectives on the problem of environmental injustice. Second, knowledge(s) of environmental injustice are grounded to different degrees in particulars and abstractions—focused on specific histories-in-place of polluting companies, on one hand, for example, and on spatial analyses of aggregated and abstracted pollution exposure data on the other. I discuss each of these tensions in turn, and suggest that in each of these arenas of debate, EJ scholars have tended to overlook the significance of the state in shaping racialized patterns of environmental injustice.

Racism versus Capitalism

Geographers' early engagement with the uneven distribution of pollution took the form of environmental equity studies that measured distributional outcomes using spatial analyses (Bowen et al 1995; Greenberg 1993; Hird 1993; McMaster et al 1997). A central debate in this literature concerned whether racial indicators or class indicators were better predictors of exposure to pollution. The so-called *race versus class debate* turned primarily on methodological choices of spatial unit(s), demographic data, and pollution data (Cutter 1995). Cumulatively, however, environmental equity studies highlighted the importance of data that had *not* been included in the spatial analyses, that is, the processes that structured the social, political and economic context in which inequitable distributions of pollution occur. As Cutter and Solecki (1996) argued:

> [t]he story of environmental justice in the southeastern United States is more complicated than simple correlations between race, income, and toxic exposures. It is a narrative that must be examined within the context of the underlying sociospatial processes that gave rise to the production of airborne toxic releases that created the riskscape. As identified here, these sociospatial processes included regional urban-industrial growth, rural underdevelopment, the structure of industrial production (small facilities versus large ones), proximity to transportation corridors, and racism. (Cutter and Solecki 1996:395)

Four aspects of this analysis stand out. First, Cutter and Solecki privilege the process of racism over the static indicators of "race" that figure in more traditional environmental equity analyses. Second, they say little about what form we should expect that racism to take. Third, racism is represented as analytically separable from capitalism, differentiated from processes which are conventionally understood to form the basis of the political economy of places and regions. Fourth, the role of the state in any of these processes, economic or otherwise, is implied but not problematized.

Some constellation of these four analytical assumptions figures prominently in the EJ scholarship discussed here. All of the work privileges racism over race *per se*. Much of the work problematizes the meaning of racism, and while most differentiates racism from capitalism, some important work links racism with the dynamics of capitalism. A common oversight, however, is that it underestimates the role of the state in shaping the meaning(s) of racialized identities that inform the EJ movement.

Early sociological analyses brought environmental racism to the center of analysis, but defined it quite narrowly. Invoking the Equal Protection Clause of the Constitution, early EJ scholar/advocates framed environmental racism as discrimination in regulatory processes such as environmental enforcement (Bullard 1993; Lavelle and Coyle 1992). Adherents of this approach implied intentional action on the part of the state and state actors, and interpreted the significance of state instrumentally through analysis of the effects of existing laws and regulations. The focus on defining racism in instrumental terms precluded attention to how racism intersects with other social relations (such as capitalism), and furthermore how the racial nature of the state might extend beyond a discrete set of policies and enforcement practices.

Legal scholars extended inquiry away from the narrowly drawn intention-based model of environmental racism. Drawing on a wider range of civil rights protections, they argued that both corporate and state actors are implicated in environmental racism on the basis of racially disparate outcomes, whether or not they act intentionally to discriminate on the basis of race (Cole 1994; Colopy 1994; Lazarus 1993). Title VI of the Civil Rights Act of 1964 figures prominently in legal analysis and highlights the fundamentally different view of racism informing this body of work. Title VI stipulates that:

> no person in the United States shall, on the ground of race, color, or national origin, be excluded from participation in, be denied the benefits of, or be subjected to discrimination under any program or activity receiving Federal financial assistance. [42 USC § 2000d (1988), cited in Colopy 1994:153]

Lazarus (1993), for example, argued that the Environmental Protection Agency (EPA) should use the fiscal leverage of federal funding under Title VI to force state environmental agencies to address the distributional impacts of their permitting processes. Colopy (1994) urged EJ activists to use both legal and regulatory dimensions of Title VI to exert pressure on the relevant federal and state government agencies. When President Clinton issued Executive Order 12898, he attached a memorandum that reminded agencies of their obligations to enforce Title VI regulations. The effect of the Title VI solution is to bring to the fore the "institutional" over the "single bad actor" model of racism (Cole 1994). Rather than assume that racial discrimination can be identified in the prejudicial behavior of individuals, and addressed on an individual basis, Title VI assumes "pervasive discrimination and segregation in American society, [and] measur[es] success by the eradication of the consequences of racism regardless of individual motive" (Colopy 1994:188).

While Title VI is only one of several remedies to environmental racism considered by legal scholars, on balance, this body of work defines racism as a systemic social phenomenon rather than a set of discrete actions (Cole 1994; Lazarus 1993). Legal scholars such as Lazarus, Colopy, Cole and others focus on evaluating legal remedies to the problem of EJ, exploring whether the state offers protections to EJ claimants or not in a given situation. In this approach, as in the sociological analyses, the workings of the state are interpreted instrumentally through analysis of what the state does and does not do; the workings of the state are not fully theorized in terms of the conditions under which the state manages racial categories in the first place.

Marxian approaches to theorizing environmental injustice maintain the analytical distinction between racism and capitalism articulated by Cutter and Solecki (1996), but focus centrally, of course, on capitalism. Political economists (Goldman 1996; Harvey 1996, 1997; Heiman 1996a, 1996b; Lake 1996) argue that conditions of environmental injustice derive structurally from the dynamics of capitalism. In this view,

[t]he fundamental problem [of environmental injustice] is of unrelenting capital accumulation and the extraordinary assymetrics of money and political power that are embedded in that process. Alternative modes of production, consumption and distribution as well as alternative modes of environmental transformation have to be explored ... This is fundamentally a class project, whether it is exactly called that or not, precisely because it entails a direct challenge to the circulation and accumulation of capital which directly dictates what environmental transformations occur and why. (Harvey 1996:401)

While opening environmental injustice to a powerful mode of analysis, the Marxist perspective on EJ activism as represented in Harvey's remarks fails to take into consideration the relationship of the EJ movement to the American civil rights movement. In doing so, it overlooks both the role of race in animating social movements and the importance of the state in shaping the claims of EJ activists. EJ activists may or may not see a necessity for struggling to form a class as understood in Marxian analysis[3] (Gilroy 1991), but it is well documented that the EJ movement originated as a form of civil rights struggle (Weiss 1996), and this legacy is important. As Omi and Winant (1994:78) powerfully demonstrate, racial movements such as the EJ movement emerge "when state institutions are thought to structure and enforce a racially unjust social order". Omi and Winant's argument points to the importance of asking whether, how, and why the state structures and enforces a racially unjust set of environmental protections, and thereby fosters or produces environmental injustice. EJ activists claim that it does; given the history of race relations in the United States, such a claim is important. It is for EJ scholars to theorize and further explore whether, why and how the state might do so.

A centrally important and distinct body of EJ scholarship rejects the analytical separation of racism and capitalism that threads through sociological, legal and Marxist/Marxian analyses, and argues that racism *cannot* be extricated from other axes of social relations (Morello-Frosch 2002; Pulido 1996, 2000). This material makes an important step in the direction of viewing race as imbricated in the apparatus of the state, because it integrates racial meanings and structures with other axes of social relations. In a critical review of EJ research, Pulido (1996) critiqued the use of static indicators of race in environmental equity studies, and argued that geographers should be investigating racism as a lived social relation rather than race as a truncated marker of social identity. She demonstrated that racism is not monolithic, nor is it aberrational; it cannot be boiled down into discrete intentional events, but is threaded through myriad social relations. She also articulated a centrally important agenda for EJ research when she argued that, rather than sidelining racism, geographic EJ scholarship should turn its attention to the racialized nature of the economy, and investigate the "range of structural, institutional and social forces which contributed to a landscape of inequality" (Pulido 1996:142).

A host of EJ scholarship in this vein identifies intersecting manifestations of *racism* and *capitalism*, such as racialized labor markets (Morello-Frosch 2002; Pulido, Sidawi and Vos 1996); job blackmail (Bullard 1990); and racialized land markets (Lord and Shutkin 1994; Pulido 2000; Pulido, Sidawi and Vos 1996). Pulido (2000) demonstrates the significance of the intersection of multiple forms of racism—white privilege and institutionalized and overt racism—for shaping an environmentally unjust metropolitan landscape. Morello-Frosch (2002) demonstrates that racism and capitalism cannot be disentangled in the racialized labor and land markets shaping US cities. This body of work powerfully demonstrates that racism is not monolithic, that it takes different forms under different circumstances, and that it is threaded through capitalist relations.

Re-entangling racism and capitalism, and tracing the effects of state policies in shaping both is an important and productive step, yet this works shies away from theorizing the role of the state in shaping the racially unjust social order that provoked EJ activism in the first place. Pulido (2000), for instance, notes the role of several state policies in her analysis of racialized land markets, including state subsidization of suburbanization, state-sanctioned redlining under the Federal Housing Act, and the rampant construction of the freeway system through numerous communities of color. She articulates a narrative familiar to urban geographers, in which state subsidization of suburbanization enabled white flight from the central city, redlining by the Home Owners Lending Corporation and the Federal Housing Authority corralled non-white homeowners into a narrow range of neighborhoods, and interstate construction disrupted and destroyed vibrant African-American and Latino communities. She discusses how these policies shaped the socio-spatial fabric of Los Angeles County, but does not move beyond their consequences to theorize the genealogy of such policies in a fundamentally racial state. Studying the racialized nature of the economy is undoubtedly important to understanding both environmental injustice and the possibilities for mitigating its effects. At the same time, as I argue below, such analyses can be deepened and enhanced by investigating the role of the state in shaping such racialized economic processes as part of a broad commitment to race management. But first, in order to lay the groundwork for thinking through the implications of the racial state for future EJ research, I continue in this section with a brief(er) consideration of two additional tensions informing EJ research: the tension between lay and expert forms of knowledge, and between the particular and the abstract as ways of knowing anything.

Lay versus Expert Forms of Knowledge

EJ activism, advocacy and scholarship alike depend in part on the production and deployment of particular forms of knowledge. Formal EJ disputes (administrative proceedings, court cases, etc) witness a persistent tension between the *lay/experiential knowledges of many EJ activists* and *expert/technocratic forms of knowledge* leveraged by state and corporate actors (Holifield 2004; Kurtz 2003). Grassroots activists are often mobilized by experience-based knowledge of persistently polluted air/water/ground quality, of acute events, of patterns of respiratory and other illnesses in their communities and so forth (Di Chiro 1992; Krauss 1998; Stein 2004). In formal administrative EJ proceedings, activists' *experiential* knowledge is confronted by state and corporate actors wielding spatially *abstract* and *expert* knowledge built on census definitions (race, class, etc) and enumeration units (Kurtz 2002, 2003). While "people of color have always thought in theoretical terms about their condition of social, political, and economic subordination" Parker and Lynn (2002:8), the abstractions of static race categories used scientifically in EJ proceedings serve to demean such lay knowledges as anecdotal and thereby diminish their relevance in the proceedings at hand. As Pulido (1996) argued, the use of static racial categories truncates consideration of the lived experience of racism; census categories of race do not tell us how such racial categories came into being and what they mean for the everyday lives of people so identified.

Koh (2002) alludes to tensions between lay and expert forms of knowledge when she highlights that "[w]hen you can't measure that which is important, you make important that which you can measure (remarks at Yale Law School, 31 May 2002), cited in Mezey 2003:1701). State agencies' technocratic approaches to managing claims of environmental injustice/racism illustrate this truism well. Unable and often unwilling to deal with the nuances of racial bias in siting decisions, environmental agencies in particular rely on spatial analyses and truncated social indicators of race and class instead. Spatial statistics conducted on aggregate enumeration units, however, are subject to the modifiable areal unit problem and other ecological fallacies and fall distressingly short of providing evidence of systematic racial discrimination in the siting of noxious facilities. Indeed spatial statistics are used in many instances to deny the existence of such racial bias at all (cf Anderton et al 1994).

It is important to take this critique further, and consider the genealogy and state sanctioning of the race variables that inform environmental equity studies and their ilk. How do these variables come to exist and to predominate in so much social science research? What assumptions do they rest upon, and what baggage do they carry? Critical race theorists remind us that the spatial-statistical evidence of racially disparate impact, or lack thereof, is itself made possible by the disciplining and regulatory power of the Census (Goldberg 2002; Harris 2000; Mezey 2003), and the Census is a tool of the racial state's effort to organize and discipline racial categories for human beings. Goldberg locates the use of race variables deep in the project of the modern racial state, which "dominate[s] through the power to categorize differentially and hierarchically, to set aside by setting apart" (2002:9). I elaborate on Goldberg's analysis and its implications for EJ scholarship below.

Particular versus Abstract Forms of Knowledge
Closely intertwined with knowledge production, a tension between the particular and the abstract pervades EJ as a political concept and fulcrum of possibility. Stated broadly, EJ activists and scholars studying EJ activism work in different ways to make the particular legible with reference to the abstract and the abstract accessible with reference to the particular. Navigating the tension between the particular and the abstract

shapes normative declarations about the possibilities for EJ activism in both Marxist/Marxian and legal scholarship.

In Harvey's illumination of this theme, he quotes Raymond Williams' critique of militant particularism (Harvey 1996:32, quoting Williams 1989:115), in which:

> because it [militant particularism] had begun as local and affirmative, assuming an unproblematic extension from its own local and community experience to a much more general movement, it was always insufficiently aware of the quite systematic obstacles which stood in the way.

As Harvey continues: "[s]uch obstacles could only be understood through abstractions capable of confronting processes not accessible to direct local experience" (p 32). Harvey appears to suggest that weak analysis on the part of EJ activists mires them in militant particularism, and by extension that pervasive racism and a racialized state apparatus are not recognized by EJ activists as systematic obstacles.

Another possibility that warrants scrutiny is that the political, legal and administrative contexts in which EJ activists make claims against the state shape the conditions of possibility for EJ activism, and that EJ activists know it. The state apparatus may well encourage a shift in focus away from abstraction and toward particularities, as in the case of hazardous waste disposal (see Lake and Disch 1992), or it may move understandings of racism away from particular intentions toward the spatial abstraction of racially disparate outcomes. In either case, the claims of EJ activists upon the state are structured in and through a tension between the particular and the abstract as manifested in law, and in civil rights law in particular. That is, the claims which are even possible for the EJ movement to make upon the state are constrained by the configuration of legal and regulatory protections. Goldberg's (2002) argument highlights that such institutional constraints derive from the structure of a state apparatus that is deeply involved in the organization and interpretation of race; the "quite systematic obstacles" faced by EJ activists include the imbrication of race and racialization in the very structure and outlook of the modern liberal state.

In each of these three tensions, then, geographers and others have acknowledged the actions of the state to some degree, but have not pursued questions about the depth and duration of the state's involvement in the very constructions of race which inform the concept of environmental injustice. Goldberg's discussion of the racial state pushes inquiry beyond these three tensions. His argument highlights that the state plays a central role in shaping racialized regional and urban economies, by using racial categories created in order to manage racialized populations. Goldberg shows that the state manages these categories and populations in pursuit of an abstracted homogeneity and liberal universality that does no small disservice to non-white populations. The implications of Goldberg's argument for these three tensions are discussed below.

Implications of the Racial State

Goldberg's (2002) analysis in *The Racial State* destabilizes the three tensions discussed above by demonstrating that to some extent they do not emerge from the contours of EJ scholarship per se, but are shaped from beyond the academy by the racial state and its effects on social and academic life. What is the racial state? Goldberg's analysis is indebted to Jessop's (1990) articulation of state theory; he views the state as neither wholly distinct from capital and civil society, but as "inherently contradictory and internally fractured, consisting not only of agencies and bureaucracies, legislatures and courts, but also of norms and principles, individuals and institutions" (Goldberg 2002:7). This view of the state assumes a complex "entanglement of identity processes, cultural and commodity flows, and state institutions, apparatuses and functions" (p 6). Such a view of the state enables investigation of the state apparatus of law that goes further than instrumental analyses of legal protections; its acknowledges that law and the rest of society are co-created, that "law constructs race at every level . . . shaping the social meanings that define races, and rendering concrete the privileges and disadvantages justified by racial ideology" (Haney Lopez 2006:xv).

That said, Goldberg is concerned with how the liberal state's commitment to homogeneity shapes racial definition, and how the state's myriad racial exclusions in the pursuit of (white) homogeneity are subsequently denied. In this sense, Goldberg's *Racial State* parallels James Scott's (1998) analysis of the state's inability to contend with and manage heterogeneity. In *Seeing Like a State,* Scott (1998) traces the efforts of modern states to impose order on chaos, to make societies legible and thereby simplify state functions of taxation, conscription and the like. Scott notes that "[r]ationalizing and standardizing what was a social hieroglyph into a legible and administratively more convenient format . . . made possible quite discriminating interventions of every kind". Scott's own interest is in the state's ordering of nature in agriculture and forestry, but the state's dependence on and forceful creation of schematic social order applies to the creation and management of racial categories of persons as well. In natural as well as human resources, the state homogenizes its vast and heterogeneous domain in order to obtain a requisite degree of control over it.

Goldberg weaves rich connections between the emergence of the modern state, and a commitment to an abstracted homogeneity enabled by liberalism and effected through racial exclusion. Focused on "the deep involvement of the state in the organization and interpretation of race" (2002:143), he examines myriad ways in which the modern liberal state "shapes race in terms of legality, threading race through law into the very fabric of the social". He notes the entwinement of racial definition through modern states from the time of Spanish colonial expansion onward; through Indian and African enslavement, seventeenth century debates over the nature of the liberal state, discussions of "national character", citizenship criteria, immigration, and colorblindness, through both apartheid and the post-apartheid movements, the emergence of "fortress Europe", and indeed the emergence of the actual fortress known as the US prison–industrial complex.

Significantly, state approaches to racial definition have shifted over the course of this history from being overt and unapologetic, to being partially obscured within liberal law. Race and the state's need to manage race to particular ends did not just disappear with the emergence of the modern liberal state. Rather, the liberal state takes as central to its purpose the task of managing racial categories in pursuit of social homogeneity, but does so through a legal apparatus which seeks to deflect attention from its work. I highlight three key moments in Goldberg's sprawling argument in order to demonstrate that the racial state also lurks behind the three central tensions in EJ scholarship outlined above.

On the Nature of Racism

How does recognizing the racial state as such inform EJ debates over the nature of environmental racism? While it has implications both for legal scholarship and for geographers' efforts to grapple with racialized regional and metropolitan economies, the focus here is on geographic work. Pulido's (2000) investigation of how the intersection of white privilege and racially discriminatory state policies shaped the socio-spatial fabric of Los Angeles County offers an instructive starting point. Pulido traces the disproportionate burden of pollution borne in communities of color to a suite of state policies, in particular, state subsidization of suburbanization, state-sanctioned redlining under the Federal Housing Act, and the construction of the freeway system. In short, these processes corralled people of color into the central city while enabling and encouraging the movement of white people to the suburbs. This is a familiar story arc; quite clearly, the state shapes urban economic processes to racialized ends. But do we leave it at that? Goldberg's work highlights that none of these policies could be enacted without the state having extensive knowledge of its population and its racial characteristics. The US state has been amassing population data using racial categories since 1820 (Haney Lopez 2006). How and why is the state interested in racial categories to begin with? What goals has the US state attempted to meet using these racial categories?

Goldberg's analysis highlights that something deeper and more hegemonic lurks behind these various policy instruments—the racial state, predicated on the construction of white homogeneity. Using the lens of the abstraction to see beyond these policy instruments to the racial state responsible for creating them, the conditions and effects of apparently disparate urban initiatives and policies come into focus as elements of a whole.

While the modern liberal state is messy, and contradictory, it has been invested in the management of racial categories since its very inception in many more arenas relevant to EJ than even Pulido identifies. Goldberg's work highlights the importance for EJ scholars of moving beyond acknowledging the effects of state policies on racialized populations to problematize the very relationship of the state to the creation and management of the racial categories that both produce the problem of environmental injustice and make it legible *as* a social problem. Insofar as EJ research focuses on the state apparatus as it exists, the lens on redressing environmental racism will be necessarily limited. As discussed below, consideration of how the racial state came to be what it is, and to follow the logics it does, takes the recognition of a racialized economy an important step further.

Lay versus Expert Knowledge: Categories of Exclusion

Both Pulido's (1996) critique of narrow conceptions of race operationalized in environmental equity studies and the lay versus expert knowledge tension more generally rest on a critique of the use of Census race variables in EJ research and practice. Goldberg demonstrates that limitations critiqued by Pulido (1996) do not emerge from the contours of research paradigms alone, but are indeed built into the fabric of society through the elaboration of the state and its legal apparatus. While the state's central role in the construction of racial categories has gone largely unremarked upon in EJ studies, critical race theorists dig deeper into the deployment of state power in the creation and use of racial categories (Goldberg 2002; Haney Lopez 1994, 2006; Harris 2000; Mezey 2003).

Concerned with such forms of classification, and with ways in which "racial definition prompted and reinforced homogeneity" (2002:5), Goldberg notes that

> central to the sorts of racial constitution that have centrally defined modernity is the power to exclude and by extension include in racially ordered terms, to dominate through the power to categorize differentially and hierarchically, to set aside by setting apart. (p 9)

State powers of categorization are probably most familiar in (but not limited to) the work of the US Census, which "refine[s] and circulate[s] the racial distinctions that mark the ever-moving boundaries of the collectively imagined nation" (Mezey 2003:1709). Mezey aptly notes the "paradoxical nature" of the US Census: "the power of classifying and counting can be aspirational, harnessed for inclusion and recognition, and it can be disciplinary, applied in ways that exclude and erase" (Mezey 2003:1713). And as critical race theorist Angela Harris notes, social classification under United States race law—"law pertaining to the formation, recognition, and maintenance of racial groups, as well as the law regulating relationships among these groups" (Harris 2000:1928)— extends well beyond census categorization to include areas of law such as "immigration, naturalization, tax, family and inheritance law, that have shaped and continue to shape American race relations" (Harris 2000:1927; see also Haney Lopez 2006 on the centrality of racial categories to immigration and naturalization proceedings).

Inherent in the law's homogenizing capacity, its exclusionary practices, and its ability to deflect critique or resistance is a dynamic in which those who resist the law's predicates are cast as outsiders, irrational, primitive, somehow beyond reason. Experiential lay knowledge of the effects of pollution that motivates so many EJ activists often resists scientific findings legitimated by positivism and its traditional embrace by the liberal state, and is hence quite often dismissed as irrational or primitive, as in the derogation of EJ activists as hysterical housewives (Kurtz 2007), and the disparagement of Latino/a activists as belonging to a primitive religion (Catholicism).

Yet marginalized groups historically struggle for inclusion in the polity, wrenching open over time the categories and abstractions on which the state is predicated (Frazer and Lacey 1993; Kofman 1995; Vasta 2000). Whereas "[l]aw's formal narrowing of heterogeneity accordingly leaves to the non-formal—the private—realm those distinctions and hierarchies that formality is unable to tolerate", EJ activists struggle to fold the distinctions of heterogeneity back into formal consideration of EJ complaints. In doing so, they do not rely entirely on a liberal conception of the state and its polity. As I argue elsewhere (Kurtz 2005), EJ activism fosters a critique of the liberal state, its reliance on abstraction and concomitant inattention to the complexity of the social, and suggests, with mixed success, civic republican and communitarian alternatives in the pursuit of EJ. In this regard, EJ activists share an important stance with critical race theorists.

Under the Cover of Abstraction

In Goldberg's argument, the erasure and exclusion effected by the Census are means to the racial state's end (goal) of homogeneity.

Homogeneity is integral to the liberal conceptions of equality and universality, the articulation of which turns on the power of legal abstraction. Goldberg argues that the modern racial state itself conjures and depends upon abstraction in the rule of law, and that this has significance for racial (in)justice. He notes that

> [i]n its imperious insistence upon universality, modern (liberal) law opposes particularity and proceeds from a stance of anonymity . . . Here law's efficiency is a function of its abstraction . . . [yet] there is a subtle sense . . . in which modern liberal legality is . . . deeply committed both to a specific ordering of social relations and to the denial of such a commitment. (2002:141)

Indeed, "[t]he proliferation of law deflects any critical or resistant social focus from the state itself, or from state authority and raw state power" (p 140). Lake and Disch's (1992) work provides an example of this dynamic; they demonstrate that environmental regulation creates openings for capital interests to produce hazardous waste, even while positing the state as a neutral arbiter between capital and civil society. Citizen action in this arena is localized; indeed citizens are obstructed from more systemic opposition. That is, liberal law as an instrument of the state both orders socio-spatial relations in particular ways and to some extent denies that it does so. Goldberg demonstrates that much the same dynamic is in place in the arena of civil rights law. The law structures claims against the state so as to permit some actions and deny others. In the civil rights arena, it does so by relying on the abstraction of homogeneity. Ironically, the abstraction which is foundational to the liberal state "makes possible civil rights protections at the same time that it makes it impossible to fully address particularities".

Goldberg's analysis of the racial state disrupts the tensions between racism and capitalism, lay and expert knowledge, and the particular and abstract as pathways to understanding environmental injustice. *Goldberg says next to nothing about EJ per se.* But key elements of his argument about how the state creates and manages race as a social relation resonate with and deepen our understanding of the environmental injustice confronted by EJ activists. His analysis points to a means to deepen theoretical understanding of environmental injustice by theorizing it as one of the myriad outcomes of a durably racialized state apparatus. The processes which foster environmental injustice require, as Harvey (1996) notes, a degree of abstraction to imagine, understand, and expose. The racial state should be investigated as one such abstraction, yet such a task is made more difficult by the tendency of the state to hide its actions, and deflect critical focus from itself. And here is where the work of Omi and Winant (1994) to theorize racial formation and the racial state can make a powerful contribution to deepening our engagement with the racial dimensions of environmental injustice. In the remainder of this chapter, I draw from critical race theory, and the theory of racial formation in particular, to suggest ways in which to reinvigorate our approach to EJ research with attention to processes of racial formation.

Rethinking Environmental In/Justice Research

Critical race theory, and specifically the work of Goldberg, and Omi and Winant, has tremendous salience for EJ research. Goldberg's (2002) elucidation of the deeply racialized nature of the modern state suggests that the state as a robust, complex and interested actor should be considered more carefully in relation to how claims of environmental injustice are articulated and acted upon by movement activists and how these are responded to by state actors. Omi and Winant focus on the twentieth century in the United States and argue that the racial state responds in limited ways to the claims of racial movements, and thus is susceptible to (limited) change over time,[4] Highlighting the potential for racially identified social movements to make at least incremental changes in the practices of the racial state, brings the dynamic interplay between the EJ movement and various state agencies into clear focus as an important point of inquiry.

Before discussing Omi and Winant's analysis, it may be useful to briefly situate their work within critical race theory (CRT). CRT focuses attention on the role of race and racism in structuring the (US) legal system, arguing that racism should not be understood as individual acts of prejudice that can be remedied by law, but as "an endemic part of American life, deeply ingrained through historical consciousness and ideological choices about race, which in turn has directly shaped the US legal system, and the ways people think about the law, racial categories and privilege" (Parker and Lynn 2002:9).

Three of the central tenets of CRT are first, "racism is a normal part of American society . . . so deeply ingrained in the political and legal structures as to be almost unrecognizable" (Saddler 2005:42). Second, CRT challenges normative social standards grounded in the experiences of white European Americans, and pointedly develops legal and other critiques of racialization based on the narratives of people of color (Bell 1992). Third, CRT critiques the basis of race law in liberalism. With its attendant color blindness, critical race theorists argue that liberalism can only address the most "egregious racial harms, ones that everyone would notice and condemn" (Delgado and Stefancic 2001:22); liberalism cannot effectively address the ordinary everyday racialization that keeps non-whites in subordinate positions in society. Critical race theorists are suspicious of the liberal concept of rights, which are more often procedural (eg right to a fair process) than substantive (eg rights to food, equal education, etc), and which, held by individuals, are considered to be alienating, distancing people from one another and effectively fragmenting social groupings (2001:22).

The task of critical race theorists "is to unveil racism in all of its variations, [but] [t]his task can be somewhat challenging because not all the forms of racism are easy to detect and identify" (Saddler 2005:42). In this regard, the concept of racial formation proffered by Omi and Winant (1994) has been deeply influential. Omi and Winant define racial formation as:

> the sociohistorical process by which racial categories are created, inhabited, transformed, and destroyed . . . [R]acial formation is a process of historically situated projects in which human bodies and social structures are represented and organized, [and is linked] to the evolution of hegemony, the way in which society is organized and ruled. (pp 55–56)

"Racialization happens as a result of racial projects [which] are . . . vastly different in scope and effect" and which range from state actions through artistic, journalistic and academic interpretations of racial conditions, to everyday individual practices. From a racial formation perspective, "race is both a matter of social structure and cultural representation [and entails] both the deep involvement of the state in the organization and interpretation of race" as well as myriad interpersonal and cultural practices (Omi and Winant 1994:5). Omi and Winant note that the state is involved in race not only as an entity that intervenes upon race relations, but also as one that is intervened upon. They note that "the state *is* inherently racial. Far from intervening in racial conflicts, the state is itself increasingly the preeminent site of racial conflict" (1994:82).

Significantly, the state is also a dynamic entity, and with respect to racial formation, Omi and Winant argue that "state institutions acquire their racial orientations from the processes of conflict with and accommodation to racially based movements" (1994:78). The state is considered to be in unstable equilibrium which is periodically disrupted and then provisionally regained in response to political pressures from across society. Racial movements like the civil rights movement, and in its wake, the EJ movement, emerge when state institutions are judged (by enough people) to be creating and enforcing a racially unjust social order. In response to political pressure from such movements, the state acts to absorb and/or insulate the racial demands being made upon it. Absorbing demands refers to adopting them in "suitably moderate form" (p 86), while insulating demands refers to confining them to largely symbolic arenas of actions. Both absorption and insulation are tools by which the state attempts to regain its equilibrium, however unstable that may be.

Omi and Winant's theorization of the racial state offers considerable insight into the origin, successes and failures of the EJ movement. It highlights the significance of the movement's origin in a racially grounded critique of the distribution of pollution. The EJ movement originated as a racial movement that voiced opposition to the state structures which fostered racially inequitable distributions of pollution. In doing so, of course, EJ activists invoked political economic processes and caught the attention of Marxian and other activists and theorists. But the movement was conceptualized from the outset as a convergence of the civil rights movement and the environmental movement (Newton 1996; Taylor 2000). Viewing the EJ movement in the historical context of post-civil rights racial activism highlights that the successes and failures of the EJ movement fit into a trajectory of activism and quiescence between the state and racial movements, and raises a series of important questions: How did the possibility arise for the EJ movement to open up the state to its claims and grievances? What have been the effects of EJ activism on various state institutions? On continuing civil rights struggle? On relationships between African-American activists and other racial minority groups?

Omi and Winant suggest the operational outlines for investigating questions such as these. They note that "[t]he [racial] state is composed of *institutions*, the *policies* they carry out, the *conditions* and *rules* which support and justify them, and the *social relations* in which they are imbedded" (1994:83, emphasis added). Pellow's (2000) work elucidating environmental inequality formation as a social-historical process (indebted to Omi and Winant) is also instructive here. Pellow notes that "reconceptualizing environmental inequality as a process changes the whole framework for theory, methodology and policy because it is difficult to explain, measure, and develop policy around a process that is not reducible to a discrete set of actions (2000:588).

Together, this work suggests that we consider processes by which the racial state fosters environmental injustice over time and in myriad contexts, as well as the processes by which the state responds to claims of EJ. Following the example set by Lake and Disch (1992), we can examine the point in a given EJ-related decision-making process in which the public is invited to participate, and then trace the longer history of those practices, in particular the play of racial categories through that history. What was the broader social context in which this set of state practices was devised? In what ways did racial categories inform the purpose of the institution and its decision-making processes? What conditions both within and beyond the state institution support and justify the categories it deploys?

Moving farther afield than environmental-decision making, as Laura Pulido urges us to do, we should "confront the underlying logics of inequity through multiple sites (Walker and Bulkeley 2006:658) with a focus on the machinations of the racial state. We can conceptualize the racial state as a stakeholder allied with other interests in the process of environmental inequity formation (Pellow 2000), and ask, what does the racial state have at stake in this process? Research in this vein could trace the ways in which the state harnesses the flow of capital to the purpose of racial exclusion, as in the case of redlining by federal mortgage authorities (HOLC and FHA). It could examine the degree to which state institutions recognize the impacts of its decisions on people of color, and the ways in which institutions respond or fail to respond to such information; how do the state's racial categories shape and constrain such responses?

Such analysis must bear in mind the shifting and multi-faceted nature of racism, and explore the ways in which the state adapts to changing understandings of the social meaning of race and racism (Omi and Winant 1994). It is worth remembering Omi and Winant's (1994:82) caveat that "[f]ar from intervening in racial conflicts, the state is itself increasingly the preeminent site of racial conflict". How is the state implicated in the shifting meanings of race in society? How do particular state institutions and practices act on understandings of race, and thereby contribute to racial projects? How are such racial projects in turn shaped by changes in the understanding of the health and environmental impacts of disamenities such as pollution? How are racialized populations understood in relation to environmental disamenities by state institutions?

Space dictates that I suggest a limited number of questions for EJ scholars to pose in relation to the racial state, but the list clearly continues. What these questions have in common is that they focus on the interactions between EJ activists and various state agencies, while problematizing the deep and durably racial character of the state. Recognizing the state's ongoing commitment to the creation and management of racial categories, under shifting social paradigms, highlights the dynamic socio-political field within which environmental injustice is recognized and acted upon as a social problem. Given that the meaning of EJ is being negotiated in the field of action between EJ activists and the state, it is important for EJ scholars to theorize and investigate the state as a robust, complex and interested actor. The rich and fruitful insights of critical race theory enable EJ scholars to do so, and to thereby deepen our understanding of the interplay between state apparatuses and institutions and social processes and social movements that delimits the possibilities for achieving EJ.

Acknowledgements

I wish to thank Ryan Holifield, Noel Castree and the anonymous reviewers for their most constructive feedback on earlier versions of this chapter. I would also like to thank Steve Holloway for inviting me to co-teach his seminar on Race and Racialization, and the very thoughtful graduate students who have enlivened discussion on this general topic. Any remaining errors or omissions are mine.

Endnotes

[1] The broad base of scholarship in this arena is conventionally referred to as the environmental justice literature, abbreviated EJ. Within that broad body of work, different attention is paid to understanding the problem of environmental injustice on the one hand and exploring possibilities for EJ as an aspiration ideal on the other. This chapter is focused on theorizing the problem of environmental injustice, but follows the convention of referring to the broader agenda as one of EJ scholarship.

[2] This chapter is by no means intended to include a comprehensive review of EJ scholarship; rather, it makes a small pass through a robust body of work to highlight particular themes that are complicated by recognition of the racial state in relation to environmental injustice and EJ struggle.

[3] Paul Gilroy observes that "class today is a contingent and necessarily indeterminate affair... Social movements centred on the experience of subordination as well as exploitation include class but are not reducible to it. Where it enables political action and organization 'race' falls into this category" (Gilroy 1991:35).

[4] While Omi and Winant are not part of the group of legal scholars who identify as critical race theorists, their theory of racial formation has been tremendously influential to critical engagements with race, and so I characterize their theory of racial formation as critical race theory in the broader sense of the term.

References

Anderton D, Anderson B, Rossi P, Oakes J, Fraser M, Webber E and Calbrese E (1994) Hazardous waste facilities: "Environmental equity" issues in metropolitan areas. *Evaluation Review* 18:123–140

Bell D (1992) *Race, Racism, and American Law*. 3rd ed. New York: Little Brown & Co

Bowen M, Salling M, Haynes K and Cyran E (1995) Toward environmental justice: Spatial equity in Ohio and Cleveland. *Annals of the Association of American Geographers* 85(4):641–663

Bullard R (ed) (1990) *Dumping in Dixie: Race, Class and Environmental Quality*. Boulder, CO: Westview Press

Bullard R (ed) (1993) *Confronting Environmental Racism: Voices from the Grassroots*. Boston: Southend Press

Cole L (1994) Environmental justice litigation: Another stone in David's sling. *Fordham Urban Law Journal* 21(3):523–545

Colopy J (1994) The road less traveled: Pursuing environmental justice through Title VI. *Stanford Environmental Law Journal* 13:125–171

Cutter S (1995) Race, class and environmental justice. *Progress in Human Geography* 19(1):111–122

Cutter S and Solecki W (1996) Setting environmental justice in space and place: Acute and chronic airborne toxic releases in the southeastern United States. *Urban Geography* 17(5):380–399

Delgado R and Stefancic J (2001) *Critical Race Theory: An Introduction*. New York: New York University Press

Di Chiro G (1992) Defining environmental justice: Women's voices and grassroots politics. *Socialist Review* 22(4):93–130

Frazer E and Lacey N (1993) *The Politics of Community*. London: Harvester Wheatsheaf

Gilroy P (1991) *There Ain't No Black in the Union Jack: The Cultural Politics of Race and Nation*. New York: Routledge

Goldberg D T (2002) *The Racial State*. Oxford: Blackwell

Goldman B (1996) What is the future of environmental justice? *Antipode* 28(2):122–143

Greenberg M (1993) Proving environmental inequity in siting locally unwanted land uses. *4 Risk–Issues in Health and Safety* 23:235–252

Haney Lopez I (1994) The social construction of race: Some observations on illusion, fabrication and choice. *Harvard Civil Rights-Civil Liberties Law Review* 29:1–62

Haeney Lopez I (2006) *White by Law: The Legal Construction of Race*. 10th anniversary ed. New York: New York University Press

Harris A (2000) Symposium on law in the twentieth century: Equality trouble: Sameness and difference in twentieth-century race law. *California Law Review* 88:1923–2015

Harvey D (1996) *Justice, Nature and the Geography of Difference*. Cambridge, MA: Blackwell

Harvey D (1997) The environment of justice. In A Merrifield and E Swyngedouw (eds) *The Urbanization of Injustice* (pp 65–99). New York: New York University Press

Heiman M (1996a) Race, waste and class: New perspectives on environmental justice. *Antipode* 28(2):11–121

Heiman M (1996b) Waste management and risk assessment: Environmental discrimination through regulation. *Urban Geography* 17(5):400–418

Hird J (1993) Environmental policy and inequity: The case of the Superfund. *Journal of Policy Analysis and Management* 12:323–343

Holifield R (2004) Neoliberalism and environmental justice in the United States environmental protection agency: Translating policy into managerial practice in hazardous waste remediation. *Geoforum* 35:285–297

Jessop B (1990) *State Theory: Putting Capitalist States in their Place*. University Park, PA: Penn State Press

Kofman E (1995) Citizenship for some but not others: Spaces of citizenship in contemporary Europe. *Political Geography* 14:121–137

Krauss C (1998) Challenging power: Toxic waste protests and the politicization of white, working-class women. In N Naples (ed.) *Community Activism and Feminist Politics: Organizing Across Race, Class and Gender* (pp 129–150). New York: Routledge

Kurtz H (2002) The politics of environmental justice as a politics of scale. In A Herod and M Wright (eds) *Geographies of Power: Placing Scale* (pp 249–273). Oxford: Blackwell

Kurtz H (2003) Scale frames and counter scale frames: Constructing the social grievance of environmental injustice. *Political Geography* 22:887–916

Kurtz H (2005) Alternative visions for citizenship practice in an environmental justice dispute. *Space and Polity* 9(1):77–91

Kurtz H (2007) Gender and environmental justice in Louisiana: Blurring the boundaries of public and private spheres. *Gender Place and Culture* 14(4):409–426

Lake R (1996) Volunteers, NIMBYs, and environmental justice: Dilemmas of democratic practice. *Antipode* 28(2):160–174

Lake R and Disch L (1992) Structural constraints and pluralist contradictions in hazardous waste regulation. *Environment and Planning A* 24:663–681

Lavelle M and Coyle M (1992) Unequal protection: The racial divide on environmental law. *The National Law Journal* 15:S1–S14

Lazarus R (1993) Pursuing "environmental justice": The distributional effects of environmental protection. *Northwest Urban Law Review* 87(3):787–857

Lord C and Shutkin W (1994) Environmental justice and the use of history. *Boston College Environmental Affairs Law Review* 22:1–26

McMaster R, Leitner H and Sheppard E (1997) GIS-based environmental equity and risk assissment: Methodological problems and prospects. *Cartography and Geographic Information Systems* 24(3):172–189

Mezey N (2003) Erasure and recognition: The Census, race and the national imagination. *Northwestern University Law Review* 97:1701–1768

Morello-Frosch R (2002) Discrimination and the political economy of environmental inequality. *Environment and Planning C–Government and Policy* 20(4):477–496

Newton D (1996) *Environmental Justice: A Reference Handbook*. Denver, CO: Island Press

Omi M and Winant H (1994) *Racial Formation in the United States: From the 1960s to the 1990s*. 2nd ed. New York: Routledge

Parker L and Lynn M (2002) What's race got to do with it? Critical race theory's conflicts with and connections to qualitative research methodology and epistemology. *Qualitative Inquiry* 8(1):7–22

Pellow D (2000) Environmental inequality formation: Toward a theory of environmental justice. *American Behavioral Scientist* 43(4):581–601

Pulido L (1996) A critical review of the methodology of environmental racism research. *Antipode* 28(2):142–159

Pulido L (2000) Rethinking environmental racism: White privilege and urban development in Southern California. *Annals of the Association of American Geographers* 90(1):12–41

Pulido L, Sidawi S and Vos R (1996) An archaeology of environmental racism in Los Angeles. *Urban Geography* 17(5):419–439

Saddler C (2005) The impact of Brown on African American students: A critical race theoretical perspective. *Educational Studies* 37(1):41–55

Sandweiss S (1998) The social construction of environmental justice. In D Camacho (ed) *Environmental Injustices, Political Struggles* (pp 31–58). Durham, NC: Duke University Press

Scott J (1998) *Seeing Like a State: How Certain Schemes to Improve the Human Condition Have Failed*. New Haven, CT: Yale University Press

Shrader-Frechette K (2002) *Environmental Justice: Creating Equality, Reclaiming Democracy*. New York: Oxford University Press US

Stein R (2004) Introduction. In R Stein (ed) *New Perspectives on Environmental Justice* (pp 1–13). New Brunswick, NJ: Rutgers University Press

Taylor D (2000) The rise of the environmental justice paradigm: Injustice framing and the social construction of environmental discourses. *American Behavioral Scientist* 43(4):508–580

Vasta E (2000) The politics of community. In E Vasta (ed) *Citizenship, Community and Democracy*. London: Macmillan Press

Walker G and Bulkeley H (2006) Editorial: Geographies of environmental justice. *Geoforum* 37(5):655–659

Part II: Spaces for Critical Environmental Justice Research

Chapter 5

Digging Deep for Justice: A Radical Re-imagination of the Artisanal Gold Mining Sector in Ghana

Petra Tschakert

Introduction

Environmental and social injustices are commonly understood as the unequal spatial distribution of pollution and toxicity and the disproportionate environmental burden born by racial minorities and poor and politically disadvantaged communities. Academic research on environmental justice became prominent in the USA in the 1980s, reflecting debates on the politics of race and civil rights. This early body of work focused on contesting the siting of polluting factories and waste disposal in predominantly black neighborhoods and on indigenous people's reservations (Agyeman 1990, 2003; Bowen et al 1995; Bullard 1990). Since the 1990s, social and environmental justice debates have refined the notion of environmental racism in the USA (Pulido 1996, 2000) and expanded in the areas of risk, social exclusion, and sustainability in the UK (Agyeman and Evans 2004; Bickerstaff and Walker 2003; Walker 1998; Walker and Bickerstaff 2000; Walker and Bulkeley 2006; Walker, Fairburn and Bickerstaff 2001).

Today, environmental justice research and campaigning are becoming more and more international. Discourses and frameworks are applied to new contexts and increasingly so in the Global South. They include inequitable distribution of hazards and risks related to air and water pollution, environmental disasters, climatic changes, food insecurity, and extractive industries. Growing inequity and environmental racism in the global realm are seen as intrinsically linked to the current neoliberal model of globalization and development. Important geographic literature exists on neoliberalism and environmental governance and quality (eg Ferguson 2006; Heyen et al 2007; Liverman 2004; McCarthy and Prudham 2004). The inequity in the new world economy, as stressed by Schlosberg (2007), lies in the unfair distribution of the profits stemming from environmentally destructive practices and the exploitation and contamination that undercut poor people's access to water, land, and other resources vital for sustaining their livelihoods.

Despite geography's vital role in analyzing and critiquing the spatialities of social and environmental injustices, there is an urgent need within the discipline to think beyond the distributive patterns of environmental risks and ills and embrace emerging concepts of the plurality of justice (Ishiyama 2003; Walker and Bulkeley 2006). Plurality in this context means also to address procedural questions of injustice. These include the social and political contexts that engender unjust outcomes as well as the recognition of the diversity of social actors, their identities, experiences, and capabilities, their rights to livelihood, and participation of the most affected in environmental decision-making and political processes (Heiman 1996; Lake 1996; McDonald 2002; Schlosberg 2003, 2004, 2007; Schroeder 2000; Walker and Eames 2006). This is particularly crucial in environmental justice scholarship outside of the Western industrialized world where pronounced inequalities of recognition are commonly tied to the nature and functioning of a postcolonial state (Williams and Mawdsley 2006).

Given this emerging emphasis on the procedural dimensions of injustice, the contributions of scholars to not only examine but also remedy the underlying causes of social-environmental injustices, theoretically and practically, are disappointingly limited. With the notable exception of recent efforts to bridge environmental justice and political ecology, focusing on the North (e.g. Brownlow 2006; Bullock and Hanna 2008; Rikoon 2006; Schroeder, St Martin and Albert 2006), few geographers have attempted to actively engage in concrete efforts to counteract the very processes that produce misrecognition and exclusion. Such a lack of action is even more deplorable as critical and radical geographers, mainly in the UK, have been convincingly arguing to go beyond the mapping of inequality and exploitation and empower and engage affected populations in solutions (Fuller and Kitchin 2004; Kitchin and Hubbard 1999; Pain 2003, 2004). Such claims appear entirely consistent with the pluralistic concept of social and environmental justice, including recognition, capabilities, agency, and participation.

This chapter attempts to fill this gap by addressing disrespect, assault, and exclusion as key elements of injustice in the artisanal (illegal) gold mining sector of Ghana—a unique case of a marginalized group creating its own injustices for which they are accused and criminalized. The challenging question that is ultimately posed is: do these illegal men and women miners deserve justice? Based on insights from critical social theory and political philosophy on the procedural dimensions of social and environmental justice, I first expose marginalization and devaluation of unregistered miners as "status injury" (Fraser 2000), a fundamental and institutionalized form of misrecognition. This novel perspective on the complex geography of environmental justice contributes to the theoretical understanding of injustice by challenging the traditional, narrow focus on inequitable spatial distribution of hazards and pollution in low income and minority communities. In the second part of the chapter, I employ the concept of "contact zones" as a procedural tool to counteract misrecognition and exclusion. Through parity-fostering participatory research and the shaping of more inclusive spaces, I propose a radical re-imagination of the artisanal gold mining sector that encourages agency and flourishing rather than alienation and flush-outs. I conclude with recommendations about the institutionalization of critical and politically committed engagement that bridge the activist–academic divide.

Theoretical Debates on the Plurality of Social and Environmental Injustice

The foundation of distributive injustice is intrinsically tied to lack of recognition. This lack of recognition is often expressed as disrespect, devaluation, insults, disenfranchisement, and oppression; it constrains and harms people and prevents them from participating and speaking for themselves (Fraser 1998, 2000; Honneth 1995, 2001; Schlosberg 2004; Taylor 1994; Young 1990). According to Young (1990), recognition is a social norm embedded in social practice and, thus, cannot be simply distributed. Fraser welcomes the complementarity of recognition and redistribution, asserting its vital centrality at a time when "an aggressively expanding capitalism is radically exacerbating economic inequality" (2000:108). She sees misrecognition as "status injury", a social and institutional subordination and injustice that precludes disrespected identities and communities from participating as peers in social life (2000, 2001). Harvey, one of the most influential geographers in this debate, stresses recognition as a key element of justice, arguing that it can only be achieved by "confronting the fundamental underlying processes (and their associated power structures, social relations, institutional configurations, discourses, and belief systems) that generate environmental and social injustices" (1996:401).

In order to counteract subordinating policies, codes, and institutionalized social practices, Fraser proposes "participatory parity" or equality of status as a central procedure to reduce distributive and recognition-based injustice. Put simply, if a person is not recognized, he or she does not participate. By the same token, if one does not participate, s/he is not recognized. For Fraser, institutionalized patterns of disrespect and disesteem—whether they occur in public education, social practices of every day interaction, or laws and policies—hamper parity and participation, as do inequities in distribution. Her participatory parity concept requires: institutional respect for cultural value and a clear change in how currently misrecognized are viewed; and adequate resources to enable participation. Along similar lines, Young (1990) argues that participatory and democratic structures, deliberation, and decision-making processes are both an element of and a condition for social justice.

There is a third element, in addition to recognition and participation, that characterizes the plurality and multiple spaces of social and environmental justice: capabilities. Following the capability approach, developed and promoted by Sen (1999, 2004, 2005) and Nussbaum (2000, 2006), attention ought to be paid to how distributed goods and bads affect people's well-being, their functioning and agency, and how they can be transformed to support the flourishing of individuals and communities. Such flourishing depends on the particularities of any given social context. Injustice occurs when this flourishing is limited, undermined, or suppressed. While Sen promotes a context-specific capability framework emphasizing individual advantages, agency, and participation, Nussbaum proposes a basic "capability set" of universal relevance. As part of her central human capabilities, she lists bodily health and integrity, control over one's environment, freedom of expression, and the right to determine one's notion of a good life. Although criticized for being prescriptive, Nussbaum's list is useful as it stresses a host of ethical dimensions that are embedded in a pluralistic concept of justice, thereby underlining moral–legal–political principles any government should assure. Sen, arguing that any list would essentially defy the plurality concept, highlights the importance of deliberative democracy and public participation in the process of eliciting particular values for well-being and social change.

Despite the multiplicity and divergence of theoretical debates about the substance and directions of environmental and social justice research, the entry points for action are straightforward and concrete. Typically, they are political struggles for justice, triggered by various forms of misrecognition, exclusion, and decimation of individual and community capabilities. I use these theoretical frameworks to explore tangible spaces of recognition, participatory parity, and flourishing in the contested artisanal mining sector in Ghana.

Environmental and Social Injustice in Extractive Industries: The Case of Ghana

In the early 1500s, King Ferdinand of Spain laid down the priorities as his conquistadors set out for the New World. "Get gold," he told them. "Humanely if possible, but at all costs, get gold." (Perlez and Johnson 2005)

Georesource extraction of oil, gas, copper, and precious minerals such as gold and diamonds appears acutely prone to social and environmental injustices, and exceedingly so in the Global South. There is ample evidence that processes of resource exploitation unequally enrich national elites and foreign corporations while leaving local communities excluded and deprived of their land-based livelihoods and their natural resources contaminated and degraded, with severe impacts on human and environmental health.

However, relatively little geographical research has been devoted to social and environmental injustices and conflicts in georesource extraction compared with a vast literature in forestry and fishing (eg McCarthy 2006; Peluso 1992; Prudham 2004; St Martin 2007). This is particularly true for Africa, with a few, albeit highly influential exceptions including the "petro-violence" in the Niger Delta (Omeje 2005; Watts 1999, 2001, 2005, 2007), diamond mining in Sierra Leone (Le Billon 2008; Richards 2001); impacts of asbestos mining among predominantly black miners (Cock and Koch 1991; McCulloch 2002), post-Apartheid environmental activism (McDonald 2002), and mercury recycling (Martinez-Alier 2001) in South Africa; and, much broader, mining as part of Africa's neoliberal era (Ferguson 2006).

The environmental and social injustices of gold mining, in particular, remain poorly understood. Yet, gold is mined in over 25 countries in Africa and in none of them without conflict. In Ghana, Africa's second largest gold producer after South Africa, there are to date roughly 40 officially recognized large-scale mining companies (Infomine 2007), including industry heavyweights AngloGold Ashanti, Newmont, Gold Fields, and Bogoso Gold Limited/Golden Star Resources (BGL) in the "Golden Triangle" in the southwestern part of the country (Figure 1). The privatization of state gold mining concerns has been boosted as part of a neoliberal economic development package beginning in the 1980s (Ferguson 2006). Rising conflicts about the availability of land are exemplified by Hilson and Yakovleva (2007) in their detailed analysis of increasingly strained relations between small- and large-scale mining.

In the case of Ghana's large-scale mining, community dislocation is one of the most severe social injustices, also reported for other African countries that promote extensive mineral exploration (Ballard and Banks 2003; Downing 2002). A total of 30,000 people were displaced during the first 5 years of Gold Fields' operations and an additional 20,000 farmers on Newmont's Ahafo mine (Owusu-Korateng, personal communication). Environmental impacts of large-scale, predominantly surface operations on surrounding communities

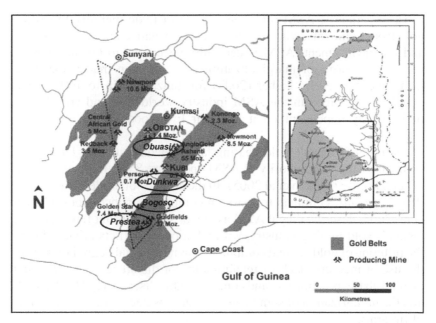

Figure 1: Gold-producing areas in Ghana, including the "Golden Triangle" (after PMI Gold Corporation; http://www.pmigoldcorp.com/i/maps/GoldenTriangle.gif). Ovals indicate research sites

include biodiversity and ecosystem loss as well as soil erosion from land clearings. Toxic chemicals, mainly cyanide, contaminate surface and groundwater, pollute the air, and leave behind tailings and waste. Environmental incidents happen regularly, including cyanide spillages in 2004 and 2006 and the bursting of a major tailing dam in 2001 (Carson et al 2005; Graham 2005; No Dirty Gold 2001). Yet, the arguably fundamental injustice in Ghana's large-scale mining sector lies in the fact that companies, many of them inter- and multinational, have been receiving preferential treatment from the state compared with small-scale, local operators, including generous tax exemptions and repatriation of profits. As in the case of postcolonial India, the state privileges the "modern" over the "traditional" and the "national" over the "local" (Gadgil and Guha 1995).

In comparison, the complexity of injustices in artisanal, small-scale mining (ASM) is poorly understood. What is well known is that Ghana is home to 300,000–500,000 people working in the ASM industry, accounting for more than 60% of the country's total mining labor force (Carson et al 2005; Hilson and Potter 2003). ASM is largely poverty driven, associated with rudimentary techniques of mineral extraction, highly manual processes, hazardous working conditions, and frequent conflicts over land and other resources (Hilson 2002).

Most importantly, ASM relies on the use of toxic mercury to extract gold from sediments, with various potential negative impacts for people and the environment. ASM is certainly not unique to Ghana, nor are the various forms of marginalization and exclusion that characterize it, as documented by Fisher (2007) for Tanzania. Up to 100 million people are engaged in the ASM industry worldwide and are directly or indirectly dependent on it for their livelihood (Veiga and Baker 2004).

As most of Ghana's gold-bearing land is demarcated for large-scale companies, the vast majority of ASM operators (85%) mine without an official license (Carson et al 2005). In the absence of a title to land, such illegal (*galamsey*) miners trespass on companies' concessions. Carson et al (2005) describe illegal ASM as one of seven conflict flashpoints in Ghana's gold mining industry, with additional conflicts arising due to the pilfering of gold ore and equipment; environmental degradation from the use of mercury; lack of rehabilitation of disturbed land; drug and alcohol abuse, prostitution, and communicable diseases (eg HIV/AIDS); and the militarization of some *galamsey* groups due to a growing inflow of firearms.

Clearly, the ASM sector in Ghana exhibits real challenges that should not be trivialized. Yet, from the perspective of social and environmental injustice, the persistent devaluation of *galamsey* gold miners constitutes an unusual case. Thousands of unregistered operators, for many of whom mining is their traditional way of live, are essentially forced into illegal and often destructive activities due to the unavailability of mining land for registration or alternative livelihoods (Hilson and Yakovleva 2007). Contrary to other communities impacted by large-scale extractive industries, *galamseyers* create their own environmental and social "bads" for which they are accused and marginalized.

Governmental and public discourse and the Ghanaian media portray *galamseyers* as a "threat", "problem", "headache", "challenge", "menace", and "violent criminals" (General News 2006; Ghana Web 2006; Mining News 2006; Palmer and Sackey 2004; PeaceFM online 2007; Regional News 2006; Ryan 2006). This anti-*galamsey* discourse continues to be promoted in spite of the state's policies to support ASM as a catalyst for poverty reduction and various attempts to regularize the sector. While these miners are portrayed as villains, the state purchases *galamsey* gold via agents licensed through the Precious Mineral Marketing Corporation. In fact, the total ASM sector in Ghana is booming, with a contribution to the national economy of estimated $ 461.1 million since 1989 (Carson et al 2005). Hence, the central question is: what constitutes the core of this injustice and can it be remedied?

Creating "Contact Zones" through Participatory Research

Social and environmental injustices often occur because of the physical distance between us and the "other"—the insulted, degraded, and excluded—and intellectual and emotional disconnects from their thoughts and perceptions. This is particularly true for socio-economically less well-off and less esteemed groups and places that are remote from our daily concerns. Without any doubt, Ghana's *galamsey* miners constitute such a group.

I argue that the fundamental discrimination that afflicts indigenous *galamsey* mining groups lies in the fact that governmental policies do *not recognize* them as equal citizens and, hence, exclude them from educational, technical, and financial support structures. Such an institutionalized form of misrecognition is what Fraser (2000) calls "status injury". It portrays *galamsey* operators as inferior, illegitimate, and dangerous. Branding them as "violent criminals" denies them the status of a full partner in safe resource extraction and decision-making and undermines good environmental stewardship. The persistent use of mercury is presented as a convenient excuse for their devaluation, yet no alternative extraction technologies are offered, and the vicious cycle of contamination and exclusion continues. At the same time, large-scale companies still use cyanide for gold extraction, self-monitor their often debatable environmental performance, and dishonor human rights. Conolly (1993) argues that persistent misrecognition, disrespect, and disempowerment are likely to fuel resentment. This is certainly the case in Ghana's ASM sector as conflicts are on the rise (Hilson 2007). Inadequate understanding of local conditions and lack of community representation fuel a highly volatile situation in an otherwise relatively peaceful nation with a long history of stability (Carson et al 2005). Creating "contact zones" as spaces for parity and participation may offer a way out of the current impasse.

Contact Zones

The term "contact zones" was originally introduced by literary scholar Pratt as "social spaces where cultures meet, clash, and grapple with each other, often in context of highly asymmetrical relations of domination and subordination" (Pratt 1992:7). While her initial focus was on cultural differences and strategies of negotiations in educational settings, the concept is now widely applied across disciplines. For instance, it also refers to hybridity of geographical locales as well as spaces, networks, and practices of resistance (Routledge 1997; Yeoh 2003). Moreover, among critical human and/or radical geographers, the concept exemplifies spaces for critical engagement and mutual learning that contest the separation of activism and the academy (Castree

1999; Merrifield 1995), similar to Routledge's "third space" (1996), encouraging scholars to interpret and effect social change.

Here, I use the concept of contact zone drawing upon both Pratt's original notion of clashing and grappling people with unequal power relations and the radical geographers' understanding of critical engagement. This blend offers a useful entry point for shaping possible spaces of participation, collaborative learning, flourishing, and co-existence between *galamsey* miners, researchers, health professionals, and governmental officials. The rationale is to employ these spaces as a conceptual and procedural tool in parity-fostering participatory research to counteract misrecognition and exclusion that characterizes Ghana's ASM sector today.

Hence, one principal aim of this research is to approach, recognize, respect, and actively involve commonly marginalized and ostracized *galamsey* miners, both men and women, in an engaging and participatory manner. It is an attempt to provide a space for expression and interaction for a subordinate group of citizens to contest daily status injuries. To avoid the paternalist trap often associated with Nussbaum's capability set (Clark 2002, 2005; Schlosberg 2007), the intention was to encourage miners to articulate their own definitions of desires, capabilities, and flourishing. In fact, Stewart (2001:1192), criticizing Nussbaum's "imposition of a set of values by an outsider", advocates more participatory approaches to engage disadvantaged groups. Building on Sen's argument to encourage public participation and discussion as well as agency as crucial elements in the process of defining elements for flourishing, Stewart argues that well-designed participatory techniques can in fact take participants beyond some of the daily constrains that limit their visions. Consequently, the concept of contact zones appears highly suitable for practicing recognition, encouraging participation, and contributing to the operationalization of the capability approach.

Research Sites

Research for this study was conducted between August 2006 and July 2007. Two *galamsey* sites were selected with the help of officials from Small-Scale Mining District Offices familiar with local artisanal operations. Both sites are located in the southwestern part of the country at the heart of Ghana's gold belt. The first site (here referred to as site 1) is an alluvial site next to Dunkwa-on-Offin (population of 45,000) in the Upper Denkyira District, Central Region. This site was located on the concession of Dunkwa Continental Goldfields, a corporation that has not been operational since 1999. The previously illegal miners are now encouraged by the Minerals Commission to register officially for a title to the concession land. The other study site (site 2) is a hard rock area next to Bogoso (population of 16,500) in the Wassa West District, Western Region. The site there is located on the northeastern edge of the 85-km long concession of Bogoso Gold Limited (now Golden Star Resources). The corporation is actively exploring and exploiting its land and has become repeatedly in conflict with infringing *galamseyers*.

In January and July 2007, the research team revisited the two sites. Both had changed significantly. Site 2 was abandoned due to Operation "Flush-Out" (a military "sweep" in November 2006) while site 1, not subjected to the crackdown, was relocated to a richer area. Nonetheless, many of the miners with whom the team had interacted the previous year were still around, either employed at new sites, operating secretly at night, or in search of alternative jobs. Several were willing to participate in a follow-up phase of the study. Participants had various socio-economic backgrounds and ranged in age from 15–42 years for women and 16–46 years for men. The number of miners at the sites fluctuated between 0 and 500. Changes reflect the availability and functioning of equipment such as excavators and water pumps as well as access regulations and controls enforced by Ghanaian and corporation security forces.

Research Methods

The qualitative and quantitative methods used in this study were integrative and to a large extent participatory. Consent was obtained from camp supervisors and individual miners. The research team started with informal interviews with seven group ("gang") leaders at the two *galamsey* sites and then conducted 41 semi-structured interviews with miners (17 men and two women) as well as non-miners (11 men and 11 women) from Dunkwa and Bogoso, two adjacent towns, to investigate advantages and disadvantages of *galamsey* work. Participating miners were chosen with the help of gang leaders while non-miners were randomly selected by approaching people with varying backgrounds in the two towns. A total of 25 informal and semi-structured interviews were conducted with people from industry, academia, citizen groups, the government, as well as miners and non-miners. The identity of some interview partners remains confidential.

Two participatory hazard maps were prepared by two groups of young miners (18–30 years) to depict dangerous and safe places on their working sites. To better understand linkages between hazards and human health, we used "body health mapping", drawing upon the work of Keith and Brophy (2004) among asbestos-exposed mine workers in Canada. Approximately 120 men and women miners participated at three sites (sites 1 and 2 and a temporary sand-digging site near Bogoso).

One volunteer at each site drew a life-size human body on a large piece of paper. One by one, the miners indicated the various body parts where they felt pain by using color-coded self-adhesive dots (gender differentiated by color). This activity was conducted anonymously at the first two sites (behind a tree and a wall, respectively), but, upon request by the participants, in front everybody else at the third site. Invited professional nurses discussed the emerging patterns of pain, likely causes, and possible solutions for healing and treatment.

In spring 2007, a total of 76 livelihood surveys were administered among miners and non-miners in the mining towns of Dunkwa, Obuasi, Japa, Bogoso, and Tarkwa to investigate livelihood preferences and choices. Participants were selected through a stratified sampling approach, involving two-thirds *galamsey* miners (33 men and 15 women) and one-third spanning the typical professions in the area, such as farming, teaching, auto mechanics, taxi drivers, carpentry, nursing, petty trade, hairdressing, and seamstresses (14 men and 14 women). Finally, we invited two groups of *galamsey* operators, seven to ten men in each group, aged 18–30 years, to depict their ideal working environment, using vision mapping. They were asked to illustrate their ideal mining site on a large sheet of paper and depict and explain every feature that would allow their flourishing as part of a visionary mining landscape. These features could include extraction and processing aspects as well as health, educational, and entertainment infrastructure.

Misrecognition and Exclusion at the Core of Ghana's Mining Conflicts

This first section of the analysis addresses misrecognition and impeded participation as so far neglected aspects of social and environmental justice in and around illegal mining sites. *Galamsey* miners operate outside the legal system, and most of them are fully aware of their infringements and other illicit actions. The key question is whether or not they are entitled to fair treatment and participatory parity.

Disrespect, Assault, and Status Injury

Misrecognition in social and environmental injustice can be defined along two conceptual lines: as individual psychological aspects of disrespect and as institutionalized practices or "status injury" (Fraser 1998). According to Honneth (1995), there are three major forms of individual disrespect, each with distinct psychological dimensions: violation of the body (torture), denial of rights, and denigration of ways of life.

In Ghana's *galamsey* sector, misrecognition in the form of psychological (and physical) disrespect is widespread. The arguably worst case is the ongoing conflict in and around Prestea Town between Bogoso Gold Ltd (BGL), property of Canadian-listed multinational Golden Star Resources, and illegal *galamsey* operators who encroach on the company's concession. The conflict, triggered by the 2002 closure of underground operations, the subsequent lay-off of 2000 local miners, and BGL's failure to relinquish uneconomic portions of the concession, is well summarized by Hilson and Yakovleva (2007). In 2005, after various vacating orders, met by strong local resistance, the military was finally called in to clear the concession land of the increasing number of *galamsey* operators (Hilson and Yakovleva 2007; member of Concerned Citizen of Prestea, personal communication 2006).

In late 2006, the government initiated Operation "Flush-Out", requested by BGL and coordinated by the Chamber of Mines (Hilson 2006). It left three people shot in Prestea and several sites emptied of both ore and equipment. The military also confiscated all equipment at mining and processing sites, burned land around existing shafts, and dug trenches to prevent *galamseyers* from returning to continue working. Nevertheless, some miners around BGL continued operating secretly and, at night, transported loads of crushed ore to processing sites 30 miles away (*galamsey* miner near Bogoso, personal communication 2006). With others, the shock of psychological and physical assaults sat too deep. "They arrest you, beat you, make you sit, then let you go at the end of the day" (*galamsey* miner from Bogoso, personal communication 2006).

It is an intrinsic lack of recognition that is at the heart of the Prestea conflict. BGL admits having a non-recognition policy with *galamsey* miners. "If we talked to them, that would mean that we recognized them as discussion partners" (BGL official, personal communication 2006). It is precisely this lack of recognition and exclusionary conduct, also reported by Jenkins and Obara (2006), that not only omits illegal operators from the company's community development work but is likely to "worsen the already volatile situation" (p 16). Calling upon the military rather than seeking dialogue—the latter being successfully demonstrated by other companies (Agyei-Twum 2006)—evokes an aura of apartheid in a company's community relations efforts. Further military sweeps of artisanal mining communities are reported on Newmont's Noyem concession in the Eastern Region (Hilson 2007) and again on BGL land, including the securing of 25 acres that *galamsey* operators had annexed for themselves (Ghana News Agency 2007a).

Intimidation of community members by soldiers and human rights abuses have also been reported (Carson et al 2005; Hilson 2007), including testimonies from Obuasi and neighboring communities where

AngloGold Ashanti (AGA), Ghana's oldest mining company, is located. According to a recent BBC report, the company has a shoot-on-site policy in the case of encroachment (Stickler 2006). Owusu-Korateng (2005), director of the Wassa Association for Communities Affected by Mining (WACAM), describes AGA's "criminal methods", including torture of *galamseyers* in private company cells. From 1994 to 1997, three *galamsey* miners were killed in this area, 16 severely beaten, and six beaten and attacked by AGA guard dogs (No Dirty Gold 2004). In 2007, AGA issued a warning to illegal miners to stay off its concessions, reporting the demolition of all illegal mining pits within the area (Ghana News Agency 2007b).

Misrecognition also undercuts political rights. The Prestea Concerned Citizens report that their right to demonstrate had been taken away and mobilizing people could end with jail (personal communication 2006). Some group members have been bought and others are threatened to silence. BGL is said to have hired "spy researchers" who proclaim apparent community satisfaction with the company over the local radio (Dumase, personal communication 2006).

Misrecognition in the form of institutionalized practices or status injury among Ghana's illegal gold miners is most likely even more widespread, albeit difficult to document. Fraser (2001) argues for a structural understanding of misrecognition more as an institutional practice rather than an individual experience. In addition to governmental policies, such status injury can occur through informally institutionalized patterns, perceptions, and customs, all portraying the misrecognized party as deficient and inferior. The notion of disrespect, being routinely disparaged in stereotypic public and cultural representations, lends itself particularly well for the case of *galamsey* miners. One example stems from a senior official of Ghana's Environmental Protection Agency: "They are illegal operators and they are armed, very violent and they don't obey any rules. They use chemicals like mercury indiscriminately. They indiscriminately mine. They mine in river bodies, in streams. I mean, anywhere" (quoted in Palmer and Sackey 2004). BGL officials describe some of the estimated 20,000 *galemseyers* on their concession as "mobile criminals from Nigeria and Côte d'Ivoire who supply firearms, drugs, and money with whom engagement is difficult" (personal communication 2006). Recently, Ghana's Deputy Minister of Lands, Forestry and Mines condemned illegal miners due to their harm to human development and the environment of mining communities (Ghana News Agency 2007c).

Finally, *galamseyers* are often poorly perceived by others in mining towns. A 54-year-old educator in Dunkwa associated alcoholism, smoking of marihuana, rapes, teenage pregnancy, HIV/AIDS, and inexistent law and order with *galamsey* miners. A 52-year-old farmer stated that "the *galamseyers* cut down cocoa trees, but often they don't pay compensation. They don't respect land; they will come and beat you up. They come from so many languages, so they are always fighting. If you have family, they can really ruin your life" (personal communication, 2006).

"If You Are Not Recognized, You Can Not Participate"
There is an obvious link between lack of respect and recognition and a person's or group's participation in the greater community, politics, and institutional order (Young 1990). Fraser (2000) asserts that both distributive inequities and institutionalized patterns of disesteem inhibit parity and participation. In Ghana's ASM sector, the misrecognized *galamsey* party has been denied full participation in state-supported educational, financial, and technical services. Participating as recognized peers means acquiring a license; yet, small-scale operators face a series of bureaucratic and procedural hurdles when trying to register (Tschakert and Singha 2007). Waiting times vary between 6 and 12 months, and first-year costs amount to 2000 Ghana Cedi (US$1900) with the District Assembly, in addition to fees with the Environmental Protection Agency. Not surprisingly, roughly 85% of people in the sector work without a license.

Institutional support for the sector has been gradually shrinking, and the seven small-scale mining district offices are notoriously under-funded (Hilson and Potter 2005). In 2006, the SSM District Office in Dunkwa pursued community level training and education, particularly with respect to mercury use for gold extraction, in fact serving five times more non-licensed miners than licensed ones. A year later, no more funding was available for mercury education programs. While the national Small-Scale Mining Association provides some technical, educational, and financial services, non-registered miners cannot be members.

Misrecognition also hampers participation in obtaining access to parts of concession lands. As expressed by a *galamsey* group in Dumase, one or two non-exploited acres would be enough to provide work for all unregistered miners in the community. Having the right to legally mine such a piece of land, even for a limited time, would put the miners on equal footing with those who have an official license. While some cooperative large-scale companies have agreed to cede parcels of unexploited land to artisanal miners, including Gold Fields Ltd who follows a "live and let live" policy (Aubynn 2006), most companies are reluctant to meet such demands (Carson et al 2005; Hilson 2007). Not having a place at the negotiation table prevents the miners from arguing for direct ceding, rather than through the Minerals Commission.

Radical Re-imagination of Illegal Gold Mining:
From Flush-Outs to Flourishing

A radical re-imagination of Ghana's artisanal mining requires first and foremost a revision of the stigmatizing *galamsey* discourse from contaminating criminals to responsible environmental stewards along with clear incentives for disenfranchised operators to register, become legal, and participate in a flourishing economic sector. From a research perspective, radical means working *with* and *for* rather than *on* these miners, not a common occurrence in the Ghanaian context. Operation "Flush-Out" and follow-up interventions that destroyed mining camps, burned land, confiscated equipment, and pushed former research participants into under-cover-of-the-night-activities constitute, I would argue, a fundamental injustice. It does so despite the fact that it involves people who themselves disrespect the law, up to half a million men and women who, one can contend, may not be entitled to fair treatment and justice. This is where the need for contact zones arises.

A "science of environmental justice", as advocated by Wing (2005:61) is "a science for the people, applied research that addresses issues of concern to communities experiencing environmental injustice, poor public health conditions, and lack of political power". In an increasingly antagonistic environment, a collaborative science that seeks to actively include ostracized miners and remedy lived injustices is a challenging undertaking. This is true for several reasons: *galamsey* miners have become increasingly suspicious of researchers because they rarely have a voice in the design and implementation of a study and hardly ever see any results; engagement with *galamsey* groups would be very difficult if not impossible without approval from and collaboration with state agents in small-scale mining district offices; yet, the same officials are often seen as a policing force rather than a support structure for unregistered operators; and engagement also means to recognize *galamseyers* as research partners; this makes the researcher suspicious in the eyes of governmental entities, large corporations, and other scientists who remain entrenched in the anti-*galamsey* discourse.

In order to create contact zones/spaces for engagement in which illegal miners could make their voices be heard and contest misrecognition and the debilitation of their capabilities, this study attempts to follow Fraser's notion of "participatory parity" (1990, 2001). Fraser argues that parity of participation is more important than inclusion per se as it requires reciprocal recognition of participants' social standing. The first step in the de-institutionalizing of parity-impeding values is to show respect. Engaging with the miners through a series of collaborative learning activities, as well as repeated visits over time and reporting back, build trust and facilitate the gradual creation of such a parity-fostering space.

The following sections summarize the three-step process employed in this research: collective learning regarding health and hazards; hearing miners' preferences for ideal working sites and alternative livelihood options; and institutionalizing recognition and engagement as a scaling up of the contact zone.

Building Participatory Parity through Collective Learning

In the past, scientific research in and around small-scale mining sites in Ghana has not been beneficial to *galamsey* operators. Not only had the miners been essentially treated as test subjects rather than active and recognized partners, especially in mercury contamination studies, but also have impact-based assessments accused them, time and again, of reckless mercury pollution without providing convincing data. The government used such scientific "proof" to tackle the "mercury problem" through retorts, despite their technical and economic disadvantages (Hilson and Pardie 2006). More worrisome, due to their notorious avoidance of community participation, effect-based studies have undermined any constructive collective learning in the sector (Tschakert and Singha 2007).

In two initial parity-fostering learning activities in this research process, we collectively explored occupational risks and human health problems encountered on *galamsey* sites. Rather than condemning miners for human and environmental risks they cause, the on-site hazard mapping encouraged them to express the risks they face on a day-by-day basis. The drawing (Figure 2) depicts a multitude of dangerous (pink) spots such as injuries from collapsing sediments and obsolete equipment, unprotected shafts in which workers can easily fall, and severe bodily pain from carrying and lifting rock ores. The few safe (green) spots are trees for resting, the lower end of sluices, and areas next to the pipes. While the research team did not have any instant solutions to these hazards at hand, the miners appreciated the interest in and recognition of their world. It actively involved them in a process that addressed their very concerns.

In the second activity, body health mapping (Figure 3), we wanted to take the notion of contact zones and applied justice research one step further. Instead of sampling their hair, blood, urine, or nails for possible mercury contamination—arguably the most common type of investigation in the ASM sector, we organized three parity-fostering learning sessions on health. They involved a professional nurse and 20–60 miners, at each of three different sites. *Galamsey* miners often don't have health insurance (Lehman 2008) and, to save money, avoid seeking professional medical advice when they experience pain or sickness. The

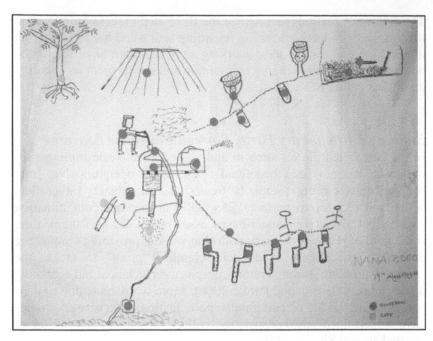

Figure 2: Participatory hazard mapping for a *galamsey* site (site 2), with dark grey dots indicating dangerous spots and light grey dots showing safe spots (fieldwork 2006)

body mapping revealed many more health problems besides mercury poisoning, commonly associated with small-scale mining. At all three sites, waist pain, chest pain, headaches, eye problems, and, at the alluvial site, foot rot collected a large number of dots. The nurses not only addressed major pains and provided simple remedies, but also demonstrated their recognition for health concerns of a usually marginalized socio-economic group. One nurse stated that now he would be more attentive to potential mining symptoms when encountered at the clinic. Having a medical expert who genuinely cared about the plights of illegal operators at an actual *galamsey* site was truly powerful. The key points of these collective learning activities have now been summarized in a health and safety manual specifically designed for small-scale miners in Ghana.[1]

What do Miners Want? "Sympathetic Imagining" and Potential Ways Out

Much of the mining discourse and literature has inherently prescriptive recommendations on how small-scale mining should occur. In Ghana, attempts to make the sector flourish have been exceedingly scarce, and

Figure 3: Body health mapping at a temporary sand digging site (next to site 2). Yellow dots illustrate pain identified by men and green dots pain identified by women (fieldwork 2006)

miners' voices in such deliberations have been even scarcer. A recent incidence illustrates *galamsey* exclusion from economic opportunities. In a 2007 news release, entitled "*Galamsey* won't qualify for PSI", the Western Regional Minister rejected a suggestion to make *galamsey* mining part of the Presidential Special Initiative[2] (Ghana News Agency 2007d). Rather than following the Peoples' Assembly's request to support the ASM sector as one enterprise project for development, the minister insisted on registration as the only means for participation. Borrowing Nussbaum's notion of "sympathetic imagining" (2006), we used vision mapping to understand what miners themselves see as key elements for flourishing. As shown below, the participating *galamsey* miners did not envision any radical changes themselves. What they want

Figure 4: Vision map depicting an ideal small-scale mining site as envisioned by a group of young miners near site 1, now operating on an officially registered parcel of land (fieldwork 2007)

is the fulfillment of basic needs to exercise artisanal mining as a viable livelihood.

The image drawn by a group of 10 young *galamseyers* (age 18–25 years) next to site 1 (Figure 4) reveals a series of structures that make sense for an alluvial mining site: a tools house to store equipment, a large pit with a well-designed road for the excavator, a machine park where the excavator can be parked and repaired, and a power plant that supplies electricity for the entire site. The miners also depicted a second large pit and set-aside land for future prospecting, which exemplifies their objective to remain on the site. In addition, the drawing shows three distinct spots for careful use and handling of mercury: a covered area where mercury is applied during the amalgamation process, a disposal area for excess mercury ("dead black"), and an area for the burning of the amalgam, represented through an open stove and two miners, one with gloves and the other with a dust mask. At present, most sites burn the mercury–gold amalgam in the open, without protection. Next to the gold processing sites is the weighing area with a scale. Also included are a canteen house, a toilet, a urinal, and a refuse dumping site. Footpaths connect the individual structures. To break the monotony of mining, the group added a poultry house that would supply additional food and diversify their work. To relax after work, the miners envision a bar right on the site ("blue spot"). At the end of the 1-hour mapping activity, and after much prompting, the *galamseyers* added a health post and an on-site ambulance. Other than the pits and the excavator, no structures on the map were actually present at the site[3] when visited in 2007. Overall, the group was truly enthusiastic to depict their ideal site; it took the participants two entire sheets of paper to capture their emerging ideas. They may have been partly stimulated by the recent initiative of the

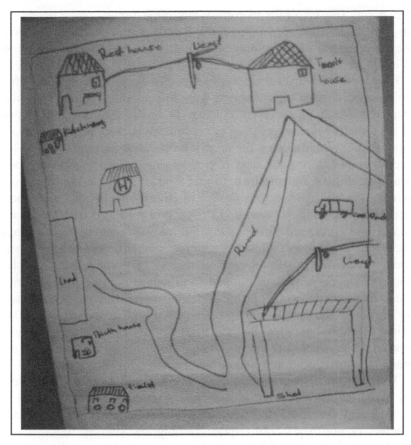

Figure 5: Vision map depicting an ideal small-scale mining site as envisioned by a group of young *galamsey* miners near site 2 whose working conditions remain uncertain (fieldwork 2007)

Dunkwa small-scale mining district office to facilitate registration of former *galamsey* operators on the now defunct concession of Dunkwa Continental Goldfields.

In contrast, the vision map produced by a group of seven *galamsey* miners near the initial site 2 is less detailed (Figure 5). Also, it was drawn within only 20 minutes. The map includes the key structures expected, albeit not necessarily the norm for hard rock mining sites: the "load" where rocks are excavated, a shed for grinding, milling, and washing, a tools house, a car park, a connection to the national electricity grid, and two lights. A strategic road links these individual elements. The group also added a rest house, a kitchen, a bath house, toilets, and a health post. The rather basic elements in the drawing mirror the limited amount of enthusiasm that these miners experienced with their profession at that point. The map fails to convey any hope for *galamsey* operators on BGL land, no sign of flourishing. It reflects the persistent military presence around illegal mining sites in that area ever since Operation "Flush-Out".

If unregistered small-scale mining receives only sparse support for flourishing, wouldn't *galamseyers* want to opt out of mining? Results from the livelihood survey reveal that 24 out of 33 male and 13 out of 15 female *galamseyers* plan to continue illegal mining for at least more than 6 months, given the relatively lucrative income. Those who intend to quit during the next 6 months cite limited prospects, poor health conditions, and a desire to further their education as main reasons. When prompted about most appealing alternative livelihoods, eight men said auto mechanic, six said taxi driver, and four said trader. Female *galamseyers* preferred batik making and trading (listed by four women each) and working as seamstresses (three women). Half of all women cited hairdresser as second best alternative to mining. Interestingly, not a single participant mentioned snail, mushroom or fish farming, grasscutter raising, or other fashionable alternative livelihoods promoted by corporations, NGOs, and governmental initiatives (Aryee 2004; Mime Consult Ltd2002). In fact, the large majority of *galamseyers* wished to be employed by one of the large mining companies. Yet, many lack the educational background required for such jobs.

To better understand the *galamsey* needs for making a living, their preferred job choices were investigated. Participants ranked four job options and described how realistic each option was for them and whether or not they would require training and start-up funds. Their needs were compared with men and women from other professions in the study areas. The results, shown in Table 1, indicate that fast income associated with a risky job (illegal mining) was clearly the least preferred option for all participants. However, for men and women *galamseyers*, this option required the least training and the lowest start-up funds. Also, given the fact that most of them are involved in high-risk operations, they judged this option as more realistic for themselves than non-miners. In contrast, regular although modest income and medium income for 20 years emerged as the miners' favorite choices, yet less realistic given training and start-up costs.

These findings suggest that the key ingredients that would make Ghana's ASM sector flourish are recognition of miners' needs, equitable distribution of mining land, and participation in the design and the control of that land. Such participation, to have operational use, needs to be contextualized at the community level (Mehrotra 2008). Involving communities in the definition of their own capabilities and desires for individual and collective functioning, as Schlosberg (2007) suggests, is not only easier but also more reasonable in a policy environment

Table 1: Comparison of income-generating preferences for miners and non-miners in Ghana, 2007

Preference of income-generating options	Male *galamsey* (*n* = 33)	Female *galamsey* (*n* = 15)	Male other profession (*n* = 14)	Female other profession (*n* = 14)
Top preference	Medium income, but for 20 years	Regular income, but not very much	Regular income, but not very much	Medium income, but for 20 years
Realistic (%)	97	87	93	86
Need training (%)	82	47	79	86
Need start-up funds (%)	48	60	71	64
Second choice	Regular income, but not very much	Medium income, but for 20 years	Medium income, but for 20 years	Regular income, but not very much
Realistic (%)	91	80	86	100
Need training (%)	70	27	86	50
Need start-up funds (%)	51	47	29	64
Third choice	High income, but only for a few years	High income, but only for a few years	High income, but only for a few years	High income, but only for a few years
Realistic (%)	70	53	93	71
Need training (%)	61	40	57	64
Need start-up funds (%)	39	13	21	28
Least desired option	Fast income, but risky job	Fast income, but risky job	Fast income, but risky job	Fast income, but risky job
Realistic (%)	94	100	79	43
Need training (%)	48	13	29	50
Need start-up funds (%)	24	7	14	14

that recognizes *galamsey* miners as equals who deserve justice through access to land. Following Sen (1999), this implies making room for human agency. In contrast, when respect and recognition are blatantly absent and individual agency undermined, no flourishing will occur.

The research also indicates that, where flourishing is unlikely, training opportunities and access to micro-credits for traditional jobs are more likely to get illegal miners out of mining than exotic drop-in projects with uncertain marketing possibilities. This substantiates earlier work by Aubynn (2004), Hilson (2007), Hilson and Banchirigah (2009), and Tschakert (2009), stressing the need for affected communities to pursue projects of their choice. In a recent attempt to provide viable capabilities for flourishing, the Ministry of Manpower, Youth and Employment now includes small-scale mining under the National Youth Employment Programme. "Youth in Mines" is a novel program within this initiative. It seeks to enable *galamsey* operators to be recognized by the Minerals Commission and to receive training to restore the environment after mining. The participants receive a monthly allowance of US$45 and, after completion of a 6-month course, the ministry provides registration and machines to get them started (*Times* 2007).

Institutionalizing Engagement

If a science of environmental justice is a science for people affected by injustices (Wing 2005), how can local parity-fostering activities and contact zones be scaled up? How can recognition, participation, and engagement be institutionalized? The state may "set an example of recognizing a socially demeaned groups and validate difference in the political realm" (Schlosberg 2007:23), but it cannot distribute recognition as it allocates, for instance, health services. Recognition requires cultural, social, and symbolic changes in the way society (dis-)regards certain groups. In order to institutionalize engagement across difference, Schlosberg (2007) advocates that recognition and representation of marginal voices are ensured, directly or through proxies, and citizen-directed policy informed by broad inclusion, ecological reflection, and social learning.

In the case of illegal gold mining in Ghana, opinions diverge to whether or not such conceptions for institutionalizing engagement are at all realistic. Some argue that *galamsey* mining is a highly lucrative undertaking that is driven by a small group of gold buyers, mercury dealers, and businessmen, often supported by affluent and influential politicians and other players outside of the actual mining areas (Nyame, personal communication 2008; Smit, personal communication 2007). Any attempt to make the sector flourish for poverty reduction and social and environmental justice may be laudable but ultimately naive. Others are more optimistic that small-scale operations will become more widespread on unused land ceded by big companies. The most encouraging example stems from Dunkwa where 12 small-scale mining companies have undergone registration since 2007 on now defunct land of former Continental Goldfields. This most recent success story demonstrates that an era of responsible mining is possible as long as incentives are provided, stigmatization is counteracted, and good practices are rewarded.

However, institutionalizing engagement for illegal artisanal mining requires digging deep for justice. While strategic public–private and advocacy partnerships have been recommended to ease the tension in Ghana's mining sector, the illegal branch, albeit in "dire need of collaborative solution and policy reform", has not been involved in any partnership (Carson et al 2005:56). To overcome existing barriers, the report argues for secured tenure of productive land, a simplified registration process, and collaboration between small- and large-scale operators. Some promising signs in addressing recognition, participation, and flourishing of the ASM sector, including *galamsey* operators, have been emerging as part of corporate social responsibility (CSR). Yet, the effectiveness of current activities has been questioned. Jenkins (2004) defines CSR in the mining industry as "balancing the diverse demands of communities, and the imperative to protect the environment, with the ever-present need to make a profit". Indeed, all national and multinational heavyweights in Ghana have been investing in physical structures for communities, mainly schools and hospitals, and community development programs, including alternative livelihood projects. Anglo Gold Ashanti's malaria prevention program has received much attention (Carson et al 2005). In 2006, Newmont was cited as the only company among the 10 largest corporations worldwide that had an over-arching CSR policy (Jenkins and Yakovleva 2006). Gifford and Kestler (2008) stress the proactive engagement of Newmont in Ghana seeking a social license to operate. The company's endeavors include a Livelihood Enhancement and Community Empowerment Program and a Social Vulnerability Forum (Engle, personal communication 2006). At the same time, local capacity building tends to fall through the cracks. Jenkins and Obara (2006) criticize insufficient consultation with communities, non-recognition policies, and deliberate exclusion of *galamsey* operators and other salient groups in community development work. Garvin et al (2008) view limited community involvement as a reflection of global processes at local scales enacted through CSR. Most critically, Hilson (2007) accuses Ghana's major mining companies of using CSR to distract from inappropriate assistance schemes, direct or indirect involvement in recent military sweeps against artisanal mining communities, massive dislocations, and human rights abuses. He argues that genuine commitment to CSR and livelihood improvement would require the relinquishing of unused portions of concession land to allow illegal operators to register and work as legal small-scale miners. So far, only one company (Gold Fields Ltd) pursues this "live and let live" policy with *galamseyers*. According to Hilson, justice is nothing more than rhetoric.

If much remains to be done to facilitate the radical re-imagination of the illegal gold mining sector at the political and institutional level, what then is the role of researchers in this process? Can we remedy some of the underlying causes of the social and environmental injustices we observe? Should we be a voice for the voiceless? I propose three options that, from a researcher's position, constitute concrete, albeit modest contributions to institutionalizing participatory parity and engagement that may also facilitate flourishing.

First, in an era of "transformative research",[4] it seems only appropriate to conceive of more creative and engaging processes that combine participatory science with deliberative processes, something Schlosberg calls a new "model for institutionalized broad inclusion and the engagement of multiple discourses, including scientific ones" (2007:198). In an attempt to scale up the initial, local contact zones and institutionalize engagement across scientific disciplines and between academia, activist community groups, miners, and governmental units, two science-practice workshops on human and environmental health and resilience in Ghana's ASM sector were held in 2008.[5] The aim was to design these workshops as participatory arenas for reflexivity, social learning, and a delicate deconstruction of difference through the involvement of a diverse range of stakeholders. They included small-scale miners whose voices, interests, and needs had been largely silenced in previous deliberative processes. Both workshops exemplified contact zones where listening and sharing formed the basis of practicable co-existence.

Second, we ought to spend more time exploring what I call the "third space of publishing" by considering appropriate local media outlets for our research findings, in addition to our obligations to the academy. For instance, one member of the Wassa Association of Communities Affected by Mining (WACAM) suggested modifying one of my academic publications for Ghana's online newspapers to a engender national debate on how to sanitize the *galamsey* industry and encourage a re-thinking among policymakers.

Third, in order to have a more direct impact on policy-influencing and implementing processes, it appears crucial to be present and, if necessary, represent the views from the margin—serve as proxy— at potentially high-impact meetings that bring together government agencies, industry, funding institutions, and the scientific community. This is not to contradict efforts to recognize artisanal miners in their own right; rather it reflects claims to bridge the gap between the realization of capabilities and functioning at the micro and the macro level (Mehrotra 2008), building on Sen's emphasis on public discussion. One such occasion was the First International Conference on Environmental Research, Technology and Policy: Building Tools and Capacity for Sustainable Production (ERTEP), held in Accra, in July 2007. While mining constituted one of the major themes and small-scale mining one specific agenda item, no representatives from the ASM sector were in attendance. If you are not recognized, you cannot participate. Our modest contribution[6] was to offer a counter-narrative to the dominant discourse of *galamsey* criminalization by demonstrating that collaborative research with illegal miners was not only possible but actually fruitful. We also shared findings on rather negligible amounts of mercury detected on *galamsey* sites. Our results were supported by similar findings from other attendants in the room. Countering the tale of *galamseyers* as reckless environmental polluters is essential for raising the profile of the sector. All in all, ERTEP provided a forum for dialogue that encompassed what Schlosberg calls "an understanding of the importance of empirical support for stakeholders and proxy positions in an institutionalized deliberative process" (2007:197).

Conclusion: Justice for Illegal Diggers is not an Oxymoron

In March 2008, the price of gold exceeded US$1000 per ounce, the highest price ever recorded in history. Over the last 18 months, the price for 1 ounce of gold in Ghana has increased from $650 to $950, a rise of 46%; and digging skyrockets. "Illegal mining in Ghana is increasingly larger in scale, they are using bulldozers . . . they have big floodlights to work at night—all of which means there is capital up front", commented Chris Anderson, Newmont's Director of Corporate and External Affairs (Ghana Review International 2008).

What are the implications of this recent gold rush for justice and flourishing? Will disadvantaged small operators become even more marginalized or can national governments ensure a more just participation in decision-making while corporate social responsibility enforces more equitable exploration? Will *galamseyers* successfully negotiate their share or is further misrecognition and exclusion likely to push the already volatile system into boundless violence and chaos? In September 2007, the World Bank manager of the Communities and Small-Scale Mining (CASM) program, Gotthard Walser, deplored the appalling conditions for growing numbers of artisanal miners around the world. He then urged for more effective partnerships between governments and mining companies in order to facilitate the flourishing of artisanal mining for economic development (World Bank 2007).

"Getting gold humanely", I argue, means to acknowledge and engage in the plurality of social and environmental justice. It means to recognize that justice has different meanings for different people in different places. In the case of illegal gold miners, in Ghana and other countries in the Global South, it implies political will to accept and remedy the underlying causes of their illegal actions. While this chapter neither endorses *galamsey* encroachment on concession land nor trivializes the sectors' real challenges—firearms, drugs, alcohol, prostitution, and land degradation *are* problems and certain sites are simply too dangerous to visit—it contends that institutionalized injustices vis-à-vis *galamseyers* have been gravely disregarded. Criminalizing and ostracizing half a million men and women not only curtails their entitlement to fair treatment but also prevents them from becoming responsible environmental stewards and participating as peers in social and economic arenas.

The core theoretical contribution of this chapter to environmental injustice debates is an explicit emphasis on recognition and participation as a counter force to criminalization. The case study explored the role of a pluralistic justice concept in rectifying institutionalized status injury at the core of discrimination. The *galamsey* case reveals a far more complex story than most orthodox scholarship on distributive justice. These illegal miners are not passive victims of hegemonic multinationals; they create their own social and environmental "bads" for which they are accused and ostracized. Yet, nothing speaks against applying the concept of justice to individuals or groups of people who, due to persistent devaluation, disenfranchisement, and exclusion from flourishing, have been essentially forced into destructive and illegal activities. According to Schlosberg (2007), the key prerequisites for recognition and participation are qualities such as interests, needs, agency, physical integrity, and the unfolding of potential. These qualities are even suggested in the context of rights debates for animals, plants, and the global atmosphere! I argue that social and environmental justice for illegal gold miners is not an oxymoron. On the contrary, it is a quest that is long overdue.

Through the use of contact zones as a conceptual and practical tool for participatory parity, the chapter contributes to bridging the bodies of environmental justice, political ecology, and radical geography, as well as operationalizing the capability approach. Contact zones allow putting Sen's (1999, 2004) notion of human agency at the center stage where it is affected people rather than local elites or external experts that decide what values and conditions for flourishing should be chosen. By drawing upon participatory research and radical geography, the focus shifts from an evaluative capability assessment to a potential capability expansion, as envisioned by Alkire (2008) and Deneulin (2008). None of the nine areas of application of the capability approach evaluated by Robyens (2006) makes use of participatory and potentially empowering methodologies. While political ecology and plurality in justice thinking pinpoint fundamental obstacles to recognition and participatory rights, it is through the practical use of contact zones that these concepts can be contextualized and operationalized at the community level. Providing such parity-fostering spaces through research does not and cannot wholly remove researchers' expectations and positionalities. Yet, I argue that spaces that allow plurality and deliberative processes to occur, as advocated by Sen (2005), from the micro to the macro level, have more legitimacy and relevance than a universal recipe for flourishing crafted from the outside.

In practice, what seems to be needed most for social and environmental justice in the small-scale mining sector is the realization of its pluralistic dimension. In addition to ceding parcels of unexploited concession land to unregistered miners and facilitating the registration process, the following actions are vital: raise the profile and status of *galamsey* miners by highlighting those who have successfully made the transition from illegal to registered operators and have become good environmental stewards; provide adequate resources that enable and sustain participation and partnerships, especially for environmentally sound technologies; and support processes that encourage flourishing in social, economic, and political realms.

The biggest obstacle for social and environmental justice in Ghana's ASM sector appears to be the fact that certain groups—politicians, chief executives, gold buyers, licensed agents, and national and international business men—not only sustain but also benefit from the *status quo*. Indeed, they have much to gain from the denigration of the many men, women, and children who work under serious human and environmental health risks and for relatively little return. While it is true that most miners live a poor existence, it seems overly simplistic to see small-scale gold mining predominantly as a mechanism for poverty reduction that, at best, requires some tweaking, without questioning the role of a small but thriving minority that perpetuates appalling injustices. Hence, the major challenge is to engross this elite group in democratic processes that enhance agency and flourishing of communities other than their own.

Contesting social, cultural, and ecological misrecognition and desolation in extractive industries, especially in the Global South, requires courage, empathy, and commitment. It requires critical engagement with remote "others" who are likely to be at the outermost ring of what Wenz (1988) calls the "concentric circle theory of environmental justice", an ethical priority for those who are closest to ourselves. At the very least, one ought to acknowledge such distant voices and the struggles they convey. Schlosberg (2007) argues that, through recognition and participation, we can counter both distance and differences. Such calls exhibit interesting parallels to recent scholarship on "geographies of care" (eg Larson 2007). From a radical geography perspective, facilitating spaces for mutual learning and engagement where academia and activism converge is both challenging and thrilling. Geographers can go beyond the mapping and disputing of spatially distributed and discursively constructed social and environmental injustices. Through parity-fostering research, collaboration, and critical engagement with multiple social actors within and beyond the industry, we can nourish capabilities and, most importantly, provide a compelling contact zone for encouraging agency and reclaiming justice.

Acknowledgements

This research was funded by a Faculty Research Grant from the Africa Research Center and a Wilson Research Initiation Grant, both from the Pennsylvania State University. I would like to acknowledge the invaluable contributions of team members Raymond Tutu, Jones Adjei, Doris Ottie-Boakye, and Iddrisu Mutaru Goro (Regional Institute for Population Studies, University of Ghana). Special thanks go to all men and women miners and professionals working in and with Ghana's mining sector who shared their stories and insights about artisanal mining. I am particularly grateful to Karl Zimmerer, James McCarthy, and Michael Bonine for their comments on earlier drafts.

Endnotes

[1] Five Penn State undergraduate students, after research in Ghana in January 2007, designed this manual as part of a course. In July 2007, a draft version was reviewed by miners and health personnel in Dunkwa and Tarkwa. Since early 2008, 1000 copies have been distributed.

[2] The President's Special Initiatives (PSIs) have been launched in five areas of activity: accelerated export development for garment and textiles, salt mining, cotton production, oil palm production, and cassava starch production. These initiatives are intended to spearhead the expansion of the economy, create jobs, and reduce poverty (especially in the rural sector).

[3] In July 2007, this site had been operational for only 6 weeks. A total of 20 gangs were working there, including six or seven women gangs. With its excavator and one motor pump, the site seemed well equipped and productive. The site's secretary was present, weighing the produced gold with a small scale and paying gang leaders in situ.

[4] "Transformative research" is the new maxim of the National Science Foundation and, as such, has swiftly altered science priorities in my own academic institution.

[5] The first workshop, "Building a Research Infrastructure on Human and Environmental Health in Ghana's Gold Mining Sector", funded by the Social Science Research Institute of the Pennsylvania State University, was held in Tarkwa, Ghana, 7–10 January 2008. A total of 22 representatives from the health sector, academia, WACAM, the Minerals Commission, and the small-scale mining sector participated. The second workshop, "US–Ghana Workshop on Resilience in Small-Scale Gold Mining", funded by NSF, was held on 7–10 July 2008 and counted diverse 39 participants. See http://www.aeseda.psu.edu/news/small-scale-mining.html

[6] My paper was presented jointly with Ghanaian and American students, their contributions being invaluable to the research.

References

Agyei-Twum F (2006) Golden Star loses 250,000oz to gold to Galamsey. *The Stateman* 28 August

Agyeman J (1990) Black people in a white landscape: Social and environmental justice. *Built Environment* 16(3):232–236

Agyeman J (2003) Constructing environmental injustice: Transatlantic tales. *Environmental Politics* 11(3):31–53

Agyeman J and Evans B (2004) "Just sustainability": The emerging discourse of environmental justice in Britain? *Geographical Journal* 170(2):155–164

Alkire S (2008) Using the capability approach: Prospective and evaluative analyses. In F Comim, M Qizilbash and S Alkire (eds) *The Capability Approach: Concepts, Measures and Applications* (pp 26–50). Cambridge, UK: Cambridge University Press

Aryee B N A (2004) Mining and sustainable development. In: Minerals Commission and Chamber of Mines, *Corporate Social Responsibility in Ghana: Extending the Frontiers of Sustainable Development*. Conference Proceedings, Western University College Tarkwa, Ghana, 2–4 September 2004

Aubynn T (2004) Enhancing opportunities in the mining communities: The ALP factor. In: Minerals Commission and Chamber of Mines, *Corporate Social Responsibility in Ghana: Extending the Frontiers of Sustainable Development*. Conference Proceedings, Western University College Tarkwa, Ghana, 2–4 September 2004

Aubynn A (2006) "Live and let live": The relationship between artisanal/small-scale and large-scale miners at Abosso Goldfields, Ghana. In G M Hilson (ed) *Small-Scale Mining, Rural Subsistence and Poverty in West Africa* (pp 227–240). Rugby, UK: ITDG Publishing

Ballard C and Banks G (2003) Resource wars: The anthropology of mining. *Annual Review of Anthropology* 52:287–313

Bickerstaff K and Walker G P (2003) The place(s) of matter: matter out of place—public understandings of air pollution. *Progress in Human Geography* 27(1):45–67.

Bowen W M, Salling K J, Haynes K E and Cyran E (1995) Towards environmental justice: Spatial equity in Ohio and Cleveland. *Annals of the Association of American Geographers* 85(4):641–663

Brownlow A (2006) An archaeology of fear and environmental change in Philadelphia. *Geoforum* 37(2):227–245

Bullard R (1990) *Dumping on Dixie: Race, Class and Environmental Quality*. Boulder, CO: Westview Press

Bullock R and Hanna K (2008) Community forestry: Mitigating or creating conflict in British Columbia? *Society & Natural Resources* 21(1):77–85

Carson M, Cottrell S, Dickman J, Gummerson E, Lee T, Miao Y, Teranishi N, Tully C and Uregian C (2005) *Managing Mineral Resources through Public-Private Partnerships: Mitigating Conflict in Ghanaian Gold Mining*. Princeton, NJ: Woodrow Wilson School of Public and International Affairs

Castree N (1999) "Out there"? "In here"? Domesticating critical geography. *Area* 31(1):81–86

Clark D A (2005) *The Capability Approach: Its Development, Critiques and Recent Advances*. GPRG-WPS-032. Global Poverty Research Group. Institute for Development Policy Management. University of Manchester, UK

Cock J and Koch E (eds) (1991) *Going Green: People, Politics, and the Environment in South Africa*. Cape Town: Oxford University Press

Conolly W (1993) *Political Theory and Modernity*. 2nd ed. Ithaca, NY: Cornell University Press

Deneulin S (2008) Beyond individual freedom and agency: structures of living together in the capability approach. In F Comim, M Qizilbash and S Alkire (eds) *The Capability Approach: Concepts, Measures and Applications* (pp 82–104). Cambridge, UK: Cambridge University Press

Downing T (2002) Avoiding new poverty: Mining-induced displacement and resettlement. Working Paper 58, Mining, Minerals and Sustainable Development

(MMSD) Project. London: International Institute for Environment and Development (IIED)

Ferguson J (2006) *Global Shadows: Africa in the Neoliberal World Order*. Durham and London: Duke University Press

Fisher E (2007) Occupying the margins: Labour integration and social exclusion in artisanal mining in Tanzania. *Development and Change* 38(4):735–760

Fraser N (1990) Rethinking the public sphere: A contribution to the critique of actually existing democracy. *Social Text* 25/26:56–80

Fraser N (1998) Social justice in the age of identity politics: Redistribution, recognition and participation. In *The Tanner Lectures on Human Values* 19. Salt Lake City, UT: University of Utah Press

Fraser N (2000) Rethinking recognition. *New Left Review* May/June:107–120

Fraser N (2001) Recognition without ethics? *Theory, Culture, and Society* 18:21–42

Fuller D and Kitchin R (2004) Radical theory/critical praxis: Academic geography beyond the academy? In D Fuller and R Kitchin (eds) *Radical Theory/Critical Praxis: Academic Geography beyond the Academy?* (pp 1–20). Praxis (e)Press: Critical work in theory and praxis

Gadgil M and Guha R (1995) *Ecology and Equity: The Use and Abuse of Nature in Contemporary India*. London and New York: Routledge

Garvin T, McGee T K, Smoyer-Tomic K E and Aubynn E A (2008). Community-company relations in gold mining in Ghana. *Journal of Environment* 90:571–586

General News (2006) "Galamsey"—now a major challenge to mining companies. 14 September

Ghana News Agency (GNA) (2007a) Security agencies swoop in illegal miners. 6 June

Ghana News Agency (GNA) (2007b) Stay off our concessions—Anglogold Ashanti. 31 May

Ghana News Agency (GNA) (2007c) Deputy Minister condemns illegal mining. 24 May

Ghana News Agency (GNA) (2007d) "Galamsey" won't qualify for PSI—Minister. 29 January

Ghana Review International (2008) Illegal mining escalates—as gold price rockets. 15 July

Ghana Web (2006) Golden Star reports government intervention to remove illegal miners from its concessions in Ghana, 2 November

Gifford B and Kestler A (2008) Toward a theory of local legitimacy by MNEs in developing nations: Newmont mining and health sustainable development in Peru. *Journal of International Management* 14(4):340–352

Graham Y (2005) Obuasi's poisoned fruit. *African Agenda* 8, 28 October

Harvey D (1996) *Justice, Nature, and the Geography of Difference*. Oxford: Blackwell

Heiman M K (1996) Race, waste and class: New perspectives on environmental justice. *Antipode* 28:111–121

Heyen N, McCarthy J, Prudham W S and Robbins P (2007) *Neoliberal Environments: False Promises and Unnatural Consequences*. London & New York: Routledge

Hilson G (2002) Small-scale mining and its socio-economic impact in developing countries. *Natural Resources Forum* 26:3–13

Hilson G (2007) Championing the rhetoric? "Corporate social responsibility" in Ghana's mining sector. *Greener Management International* 53:43–56

Hilson G and Banchirigah S M (2009) Are alternative livelihood projects alleviating poverty in mining communities? Experiences from Ghana. *Journal of Development Studies* 45(2):172–196

Hilson G and Pardie S (2006) Mercury: An agent of poverty in Ghana's small-scale gold-mining sector? *Resources Policy* 31:106–116

Hilson G and Potter C (2003) Why is illegal gold mining activity so ubiquitous throughout rural Ghana? *African Development Revue* 15(2):237–270

Hilson G and Potter C (2005) Structural adjustment and subsistence industry: Artisanal gold mining in Ghana. *Development and Change* 36 (1):103–131

Hilson G and Yakovleva N (2007) Strained relations: A critical analysis of the mining conflict in Prestea, Ghana. *Political Geography* 26(1):1–22

Honneth A (1995) *The Struggle for Recognition: The Moral Grammar of Social Conflicts*. Cambridge, MA: MIT Press

Honneth A (2001) Recognition or redistribution? Changing perspectives on the moral order of society. *Theory, Culture, and Nature* 18(203):43–55

Infomine (2007) Ghana mining companies. 23 November. http://www.infomine.com/countries/ghana.asp Accessed 23 December 2008

Ishiyama N (2003) Environmental justice and American Indian tribal sovereignty: Case study of a land-use conflict in Skull Valley, Utah. *Antipode* 35(1):119–139

Jenkins H (2004) Corporate social responsibility and the mining industry: Conflicts and constructs. *Corporate Social Responsibility and Environmental Management* 11:23–24

Jenkins H and Obara L (2006) "Corporate social responsibility (CSR) in the mining industry—the risk of community dependence." Paper presented at The Corporate Responsibility Research Conference, Dublin, 4–5 September

Jenkins H and Yakovleva N (2006) Corporate social responsibility in the mining industry: Exploring trends in social and environmental disclosure. *Journal of Cleaner Production* 14:271–284

Keith M M and Brophy J T (2004) Participatory mapping of occupational hazards and disease among asbestos-exposed workers from a foundry and insulation complex in Canada. *International Journal of Environmental Health* 10:144–153

Kitchin R K and Hubbard P J (1999) Research, action and "critical" geographies. *Area* 31(3):195–198

Lake L R W (1996) Volunteers, NIMBYs, and environmental justice: Dilemmas of democratic practice. *Antipode* 28:160–174

Larson V (2007) Geographies of care and responsibility. *Annals of the Association of American Geographers* 97(1):1–11

Le Billon P (2008) Diamond wars? Conflict diamonds and geographies of resource wars. *Annals of the Association of American Geographers* 98(2):345–372

Lehman J S (2008) "Inadequate access to social services as an element of marginalization: The case of health insurance for small-scale gold miners in Ghana." Unpublished Honors thesis, Pennsylvania State University

Liverman D M (2004) Who governs, at what scale and at what price? Geography, environmental governance and the commodification of nature. *Annals of the Association of American Geographers* 94(4):734–738

Martinez-Alier J (2001) *The Environmentalism of the Poor: A Study of Ecological Conflicts and Valuation*. Cheltenham, UK: Edward Elgar

Mate K (1999) Boom in Ghana's golden enclave. UN Office of Communications and Public Information. http://www.un.org/ecosocdev/geninfo/afrec/subjindx/113ghana4.htm Accessed 27 July 2009

McCarthy J (2006) Neoliberalism and the politics of alternatives: Community forestry in British Columbia and the United States. *Annals of the Association of American Geographers* 96(1):84–104

McCarthy J and Prudham S (2004) Neoliberal nature and the nature of neoliberalism. *Geoforum* 35(3):275–283

McCulloch J (2002) *Asbestos Blues: Labour, Capital, Physicians and the State in South Africa.* African Issues Series. Oxford: James Curry, and Indianapolis: Indiana University Press

McDonald D A (ed) (2002) *Environmental Justice in South Africa.* Athens, OH: Ohio University Press

Mehrotra S (2008) Democracy, decentralization and access to basic services: An elaboration on Sen's capability approach. In F Comim, M Qizilbash and S Alkire (eds) *The Capability Approach: Concepts, Measures and Applications* (pp 385–420). Cambridge, UK: Cambridge University Press

Merrifield A (1995) Situated knowledge through exploration: Reflections on Bunge's "Geographical Expeditions". *Antipode* 27 (1):49–70

Mime Consult Ltd (2002) *Poverty Eradication and Sustainable Livelihoods: Focusing on Artisanal Mining Communities. Ghana.* Final report. Prepared for UNDP/UNDESA, RAF 99/023. Accra, Ghana

Mining News (2006) Small-scale mines. The Ghana Chamber of Mines. 2(2), June: 6

No Dirty Gold (2001) Wassa District, Ghana. http://www.nodirtygold.org/wassa_district_ghana.cfm Accessed 27 July 2009

No Dirty Gold (2004) Sansu, Ghana. http://www.nodirtygold.org/sansu_ghana.cfm Accessed 27 July 2009

Nussbaum M C (2000) *Women and Human Development: The Capabilities Approach.* Oxford: Oxford University Press

Nussbaum M C (2006) *Frontiers of Justice: Disability, Nationality, Species Membership.* Cambridge, MA: Harvard University Press

Omeje K (2005) Oil conflict in Nigeria: Contending issues and perspectives of the local Niger Delta people. *New Political Economy* 10(3):321–334

Owusu-Korateng D (2005) WACAM condemns the shooting of small scale miners by Anglogold Ashanti, Obuasi Mine. *WACAM Press Release* 29 June

Pain R (2003) Social geography: An action-oriented approach. *Progress in Human Geography* 25(5):649–657

Pain R (2004) Social geography: Participatory research. *Progress in Human Geography* 28(5):652–663

Palmer K and Sackey S (2004) Illegal mining a threat in Ghana. Ghana Web 10 October

PeaceFM Online (2007) Galamsey operators re-surface. 20 April

Peluso N L (1992) *Rich Forests, Poor People: Resource Control and Resistance in Java.* Berkeley, CA: University of California Press

Perlez J and Johnson K (2005) Behind gold's glitter: Torn lands and pointed questions. *New York Times* 24 October

Pratt M L (1992) *Imperial Eyes: Travel Writing and Transculturation.* London & New York: Routledge

Prudham S (2004) *Knock on Wood: Nature as Commodity in Douglas-Fir Country.* New York: Routledge

Pulido L (1996) A critical review of the methodology of environmental racism research. *Antipode* 28:142–159

Pulido L (2000) Rethinking environmental racism: White privilege and urban development in Southern California. *Annals of the Association of American Geographers* 90:12–40

Regional News (2006) Illegal mining destroying farmlands in Talensi-Nab-Dam District. 10 November

Richards P (2001) Are "forest" wars in Africa resource conflicts? The case of Sierra Leone. In N L Peluso and M Watts (eds) *Violent Environments* (pp 65–82). Ithaca, NY: Cornell University Press

Rikoon J S (2006) Wild horses and the political ecology of nature restoration in the Missouri Ozarks. *Geoforum* 37(2):200–211

Robyens I (2006) The capability approach in practice. *The Journal of Political Philosophy* 14(3):351–376

Routledge P (1996) The third space as critical engagement. *Antipode* 28:399–419

Routledge P (1997) A spatiality of resistances: Theory and practice in Nepal's revolution of 1990. In S Pile and M Keith (eds) *Geographies of Resistance* (pp 68–86). London: Routledge

Ryan O (2006) Ghana gold diggers pose headache for mining firms. *Reuters* 20 May

Schlosberg D (2003) The justice of environmental justice: Reconciling equity, recognition, and participation in a political movement. In A Light and A de-Shalit (eds) *Moral and Practical Reasoning in Environmental Practices* (pp 77–106). Cambridge, MA: MIT Press

Schlosberg D (2004) Reconceiving environmental justice: Global movements and political theories. *Environmental Politics* 13(3):517–540

Schlosberg D (2007) *Defining Environmental Justice: Theories, Movements, and Nature*. Oxford: Oxford University Press

Schroeder R (2000) Beyond distributive justice: Resource extraction and environmental justice in the tropics. In C Zerner (ed) *People, Plants and Justice: The Politics of Nature Conservation* (pp 52–66). New York: Columbia University Press

Schroeder R A, St Martin K and Albert K E (2006) Political ecology in North America: Discovering the Third World within? *Geoforum* 37(2):163–168

Sen A (1999) *Commodities and Capabilities*. Oxford: Oxford University Press

Sen A (2004) Capabilities, lists and public reason: Continuing the conversation. *Feminist Economics* 10(3):77–80

Sen A (2005) Human rights and capabilities. *Journal of Human Development* 6(2):151–166

St Martin K (2007) The difference that class makes: Neoliberalization and non-capitalism in the fishing industry of New England. *Antipode* 39(3):527–549

Stewart F (2001) Women and human development: The capabilities approach [book review]. *Journal of International Development* 13:1189–1202

Stickler A (2006) Ghana's ruthless corporate gold rush. BBC File On 4, 18 July. http://www.bbc.co.uk/radio/aod/news.shtml?radio4/fileon4 Accessed 19 December 2008

Taylor C (1994) *Multiculturalism*, A Gutman (ed). Princeton, NJ: Princeton University Press

Times (2007) "Galamsey" to be part of NYEP. 18 April

Tschakert P (2009) Recognizing and nurturing artisanal mining as a viable livelihood. *Resources Policy* 34(1–2):24–31

Tschakert P and Singha K (2007) Contaminated identities: Mercury and marginalization in Ghana's artisanal mining sector. *Geoforum* 38:1304–1321

Veiga M M and Baker R (2004) *Protocols for Environmental and Health Assessment of Mercury Released by Artisanal and Small Scale Miners*. Report to the Global Mercury Project: Removal of barriers to introduction of cleaner artisanal gold mining and extraction technologies, GEF/UNDP/UNIDO. http://www.globalmercury.org Accessed 23 December 2008

Walker G P (1998) Environmental justice and the politics of risk. *Town and Country Planning* 67(11):358–359

Walker G P and Bickerstaff K (2000) Polluting the poor: An emerging environmental justice agenda for the UK? *Critical Urban Studies: Occasional Papers*. Centre for Urban and Community Research, Goldsmiths College, University of London

Walker G P and Bulkeley H (2006) Geographies of environmental justice [editorial].
    Geoforum 37:655–659
Walker G P and Eames M (2006) *Environmental inequalities*. Cross-cutting themes
    for the ESRC/NERC transdisciplinary seminar series on environmental inequalities
    2006–8
Walker G P, Fairburn J and Bickerstaff K (2001) Ethnicity and risk: The characteristics of
    populations in census wards containing major accident hazards in England and Wales.
    Occasional Papers, Series A: Geographical Research. Department of Geography,
    Staffordshire University
Watts M (1999) Petro-violence: Some thoughts on community, extraction, and political
    ecology. Working Paper 99-I. Institute of International Studies, University of Berkeley
Watts M (2001) Petro-violence: Community, extraction, and political ecology of a
    mythic commodity. In N Peluso and M Watts (eds) *Violent Environments* (pp 189–
    212). Ithaca: Cornell University Press
Watts M (2005) Righteous oil? Human rights, the oil complex and corporate social
    responsibility. *Annual Review of Environment and Resources* 30:9.1–9.35
Watts M (2007) Oil inferno. Crisis in Nigeria. *CounterPunch* 2 January.
    http://www.counterpunch.org/watts01022006.html Accessed 23 December 2008
Wenz P (1988) *Environmental Justice*. Albany, NY: SUNY Press
Williams G and Mawdsley E (2006) Postcolonial environmental justice: Government
    and governance in India. *Geoforum* 37:660–670
Wing S (2005) Environmental justice, science, and public health. Environmental Health
    Perspectives. *Essays on the Future of Environmental Health Research* 54–62
World Bank (2007) Appalling conditions for growing number of artisanal miners, says
    World Bank's CASM Initiative. Press release no 2007/059/SDN, 5 September
Yeoh B S A (2003) *Contesting Space in Colonial Singapore: Power Relations in the
    Urban Build Environment*. Singapore: Singapore University Press
Young I M (1990) *Justice and the Politics of Difference*. Princeton, NJ: Princeton
    University Press

Chapter 6

Benevolent and Benign? Using Environmental Justice to Investigate Waste-related Impacts of Ecotourism in Destination Communities

Zoë A. Meletis and Lisa M. Campbell

Consumption and production patterns, especially in nations with wasteful "throw-away" lifestyles like the United States, and the interests of transnational corporations create and maintain unequal and unjust waste burdens within and between affluent and poor communities, states, and regions of the world. (Bullard and Johnson 2000:572)

Introduction

Why Environmental Justice?
In 2002, my co-author Lisa Campbell and I traveled to Tortuguero, Costa Rica, to assess its potential as a field site for my dissertation research on the solid waste impacts of ecotourism. Lisa had worked in Tortuguero and knew that the community had solid waste management problems. While I was initially interested in documenting solid waste generation associated with ecotourism (eg through waste-stream analysis), it quickly became apparent that the waste problems were more than technical. During this initial visit, the local "recycling plant" was closed for long periods of time and garbage was being left outside the plant, overflowing from local waste receptacles, and being burned and dumped around the village and on the beach. The plant had closed after several businesses and households had refused to pay for services, and the plant's ATV had been stolen. Local politics pitted the women's organization in charge of running the recycling plant at that time against a local entrepreneur proposing alternatives, and his supporters. In this atmosphere, my interest in measuring waste was soon displaced by the

need to understand the politics of the waste crisis and how this related to ecotourism, a supposedly green form of development.

In considering the nature of the waste crisis in Tortuguero, I found myself drawn to the literature on environmental justice (EJ). The siting of waste facilities has traditionally been a predominant concern in EJ, and I saw some striking similarities between the situation in Tortuguero and waste-related conflicts described in the EJ literature. First, the "recycling plant" (waste treatment and storage plant) in Tortuguero is located in the center of the village where local people live, work, and play, and far from the major hotels/lodges that generate much of the waste; there is a spatial aspect of inequity in waste generation, disposal, and potential related impacts.

Second, the majority of tourists who visit Tortuguero are North American or European and the majority of lodge owners are North American and European expatriates or non-resident Costa Ricans. In contrast, the village is mostly composed of Afro-Caribbean residents and more recent Latino immigrants from Costa Rica and Nicaragua; the waste crisis therefore includes class and race-related inequities.

Third, even when the recycling plant (hereafter the plant) was functioning at its best, the potential for negative environmental and health impacts for those living in proximity to it was evident. While most of these impacts have not been formally measured, concern was high enough to motivate the director of the shared regional health clinic to file an official complaint (*denuncia*) against the regional municipality of Pococí for neglecting its duties in Tortuguero (personal communication with various respondents 2004).

Finally, local residents used the language of justice to describe the waste crisis. They saw the lodges and other powerful tourism actors as greatly benefiting from tourism to Tortuguero, while residents were bearing heavy costs related to its increasing waste burden. One respondent said this about the waste crisis: "The hotels create a lot of plastic, etc. The municipality doesn't offer support to get the garbage out to La Pavona. The whole system has to change, including the administration of the plant..." (R53 2004).[1] Ray Hooker, a local respondent who demanded to be cited in association with his statements, described the waste crisis in the following way, in 2004:

> The biggest impact has to be garbage because here—there is nothing we can do with it. It's very difficult to get garbage out of here, and the amount of garbage is growing. Tourists don't throw garbage but they [produce] it, and the hotels don't [do their part]. The recycling doesn't work and there is no transportation to get garbage out of here. Garbage is harder to deal with here ... So I think that maybe it's very bad and it's also very urgent. I hope that it gets better because if it doesn't, we

will lose tourism and without tourism, we go back to zero. Tourism is the only revenue source here, we can't lie to ourselves.

For the purposes of this chapter, we adopt Pulido's definition of EJ. We consider it to be a "broad set of concerns . . . focused on the relationship between marginalized groups and environmental issues, including the elitism of mainstream environmentalism . . . , the biased nature of environmental policy . . . , the disproportionate exposure of marginalized peoples to polluted environments, and the limited participation of marginalized peoples in environmental affairs" (Pulido 1996:142, citing Taylor 1992). In this chapter, we explore how Tortuguero's isolation and marginalization influence the local solid waste crisis and we discuss the geographic inequity (Bullard 1994) of where waste is generated in Tortuguero and where waste-related impacts occur. We combine research in Tortuguero with relevant literature to explore the utility of EJ for examining one type of environmental impact associated with ecotourism, and, by extension, for considering justice issues associated with ecotourism more generally. In Tortuguero, the waste crisis is both constitutive and reflective of problems with ecotourism.

Although both the EJ movement and related scholarship originated in the USA and these roots remain strong, EJ scholarship is increasingly applied both in international contexts and to *global* environmental issues. Authors and activists have expanded the concept of environmental injustices beyond the traditional focus on race-related injustices, to consider environmental conflicts in different types of marginalized communities around the world (see eg Adeola 2000; Belkhir and Adeola 1997; Carruthers 2008; Kuletz 2002; McDonald 2002; Pellow and Brulle 2005). The rationale for expanding EJ scholarship and activism is strong: "If we examine environmental issues internationally, the same domestic pattern of disproportionate exposure to environmental hazards and degradation exists worldwide among those who are nonwhite, poor, less educated, and politically less powerful" (Alston and Brown 1993:179).

Despite the expansion of EJ, some industries and their impacts remain largely absent from the literature. In this chapter, we contribute to the growing scholarship considering the diversification of EJ by framing a solid waste crisis in an unexpected EJ context: the ecotourism-based community of Tortuguero, Costa Rica. Although Tortuguero appears to have little in common with many sites of historic EJ struggles (eg is it neither urban nor industrialized; local residents gain substantive financial benefits from the industry in question), we suggest Tortuguero shares characteristics with communities struggling with more conventional EJ conflicts. We explore these characteristics to frame the solid waste crisis in Tortuguero and illustrate the similarities between communities experiencing more conventional EJ issues and isolated ecotourism-based communities in the Global South. In doing so, we demonstrate three things. First, despite largely being absent from the EJ scholarship, we suggest that service industries like tourism/ecotourism can create environmental injustices that disproportionately impact communities that are marginalized in certain ways. Second, we highlight the roles that geographical location and socio-economic, cultural, technical, and political marginality often play in both siting ecotourism development and complicating environmental impact management. Third, building on the work of others (eg Floyd and Johnson 2002; Peña 2005; Porter and Tarrant 2001; Zebich-Knos 2008), we illustrate the utility of introducing EJ concepts and theory into the study of tourism. Finally, we consider the ways in which the Tortuguero case is different than those conventionally treated in EJ, and consider what this means both for Tortuguero and for EJ theory.

In describing the case of Tortuguero, Costa Rica, we draw primarily on research conducted by Meletis (2002–2004). During much of this time period, inadequate solid waste management created undesirable impacts visible on the landscape, and was considered a "crisis" by many local residents and organizations, and the director of the regional health clinic (Meletis 2007). Meletis conducted over 70 interviews with residents[2] of Tortuguero and 1001 surveys of tourists visiting the area, and witnessed a waste audit at the recycling plant (Camacho 2003). While full analyses of the field data are presented elsewhere (Meletis 2007), we draw upon results of this work, additional fieldwork observations, and other research in Tortuguero.

Environmental Justice

Waste-related Conflicts and Environmental Justice

The siting of waste facilities is one of the earliest issues addressed in the EJ movement and literature. Key examples include: the case of *Bean vs SouthWestern Waste Management* (1979) that brought the first charge of discrimination in waste facility siting under civil rights law (Bullard 2001); analysis of the spatial distribution of solid waste disposal sites in the USA in the 1970s that revealed race as the determinant in siting (Bullard 1993); the mobilization of large numbers of African-Americans against the planned siting of a PCB contaminated landfill site in Warren County, North Carolina, in 1982 (Floyd and Johnson 2002); a study prompted by the Warren County protests, which revealed that three-quarters of commercial hazardous waste landfills in eight southern states were in predominantly African-American neighborhoods (US General Accounting Office 1983; Bullard

2001); and Robert Bullard's seminal work *Dumping in Dixie. Race, Class and Environmental Quality* (1990). These cases influenced policy with the introduction of federally mandated procedures for addressing EJ (Simpson 2002) and a federal EJ office (Bullard and Johnson 2000). In spite of these accomplishments, waste-related injustices and their relationships with marginalized communities remain an EJ topic of concern.

EJ is now being used to frame waste-related injustices in an international context. In 1991, the Global North's practice of using the Global South as a "dumping ground" for unwanted wastes gained attention after a memo from then chief World Bank economist Lawrence Summers "encouraging" more waste trade to the Global South was publicly leaked in order to draw attention to the issue (Greenpeace 1992, cited in Bullard and Johnson 2000). EJ analyses of international and global waste injustices include: Wu and Wang's (2002) analysis of foreign hazardous wastes imports into China from a human rights perspective; Njeru's (2006) combination of political ecology and environmental justice to explain plastic-bag related waste in Nairobi; and Moore's (2008) work on *colonia*-generated waste crises in Mexico as evidence of environmental injustices and related activism in Latin America.

As the geographic scope of EJ scholarship and activism has broadened, so has the range of issues considered. In the literature, EJ is now used to address mining (Halder 2003), nuclear contamination (Kuletz 2002), oil production (Comfort 2002), pesticide exposure (Peisch no date), access to environmental goods (Walker and Bulkeley 2006), income inequalities, housing inequalities, homelessness, access to services, transportation issues, the redevelopment of brownfields (Lee 1996, cited in Schlosberg 1999), inequities in access to urban greenspace (Heynen, Perkins and Roy 2008; Perkins and Heynen 2004),[3] occupational health and safety issues, and the provision of basic human rights (eg Goldman 1996).

Tourism and Environmental Justice

Despite the expansion of EJ, some industries and their impacts have escaped the attention of EJ scholars and activists. Tourism and service industries in general are underrepresented both in terms of active, organized EJ struggles on the ground and coverage in the literature. For example, Benford's table of 52 EJ-movement-associated issues includes "parks and recreation" but not tourism (Benford 2005).

Tourism researchers are not unaware of justice issues, however. The politics of international tourism development have been topics of interest since at least the 1970s (see eg Alexander and McGregor 2000;

Bhattarai, Conway and Shrestha 2005; Britton 1982; Brockington 2004; Duffy 2002; Hall 2003; Homewood 1991; Hughes 2005; Trist 1999). Scholars have explored demographic and standard-of-living differences between "hosts" and "guests" (eg Torres and Skillicorn 2004), and the links between tourism and colonialism (Hall 1994; Mowforth and Munt 1998), among other issues. An extensive literature detailing the negative environmental impacts of tourism also dates to the 1970s (eg Brown et al 1997; Cohen 1978; Fauzi and Buchary 2002; Hillery et al 2001; Lindberg, Enriquez and Sproule 1996; Orams 1999; Weaver 1999; Weaver 2002), and addresses issues like tourism-generated solid waste problems experienced in small island tourism destinations (Gregory 1999; Kirby 2004) and parks and protected areas (eg Kuo and Yu 2001; Stern et al 2003). A substantial "parks and peoples" literature is fundamentally concerned with justice issues associated with dislocation of peoples from park-appropriated lands, restrictions imposed on local behaviors and access to resources, actions and policies privileging tourist access, and local responses to such impositions (Akama 1996; Campbell, Gray and Meletis 2008; Carrier and Macleod 2005; Charnley 2005; Geisler 2003; Johnston 2003; Sindiga 1995; West et al 2003). Furthermore, tourism-related activism continues to call attention to tourism's undesirable environmental impacts and the uneven distribution of tourism costs and benefits, especially in isolated regions of the Global South (see www.tourismconcern.org.uk/). For example, EQUATIONS, an organization in India, recently distributed a "call for action" entitled "No more holidays from accountability!" It emphasizes the displacement of traditional forest dwellers, repeated relaxations of "no development zones" in coastal areas, and tourism's exemption from the Environmental Impact Assessment Notification process among their concerns, and demands that the tourism and travel industry "walk their talk on their commitment to the environment" (EQUATIONS 2008).

In spite of longstanding concerns regarding injustices created by tourism development and its secondary impacts, tourism scholars and activists have rarely explicitly incorporated EJ into their works. Similarly, few EJ scholars address tourism as a standalone topic. Khan (2002), for instance, mentions ecotourism in South Africa, but focuses on the larger context of conservation as a historically colonial enterprise that "was characterized by a wildlife-centered, preservationist approach which appealed mainly to the affluent, educated, and largely white minority" (Khan 2002:15). Cock and Fig (2002) go a step further and detail struggles to correct racism-driven inequities in access to economic opportunities, training, cultural renewal, and alternative resource economies associated with park-based tourism development. They also provide examples of access-related changes that have helped to re-balance the distribution of some tourism-related opportunities (eg allowing and supporting artists to sell their wares to tourists in Kruger National Park; establishing programs to integrate small-scale local agriculture into tourism-fed restaurants and shops).

Some explicit applications of EJ concepts to tourism have focused on access to recreational resources such as national parks and beaches by marginalized community members (Floyd and Johnson 2002; Garcia 2002; Tarrant and Cordell 1999). Porter and Tarrant's (2001) study, for instance, considered negative tourism-related impacts on American low-income, blue collar neighborhoods with federal tourism sites nearby. They also discuss populations relocated for the creation of American parks and highlight research on minority groups' exposures to contaminated fish through recreational fishing (Burger et al 1999, Toth and Brown 1997, West 1992, and West et al 1992, all cited in Floyd and Johnson 2002). The selective re-construction, gentrification and potential "Disneyfication" of the French Quarter in New Orleans to serve tourism is also being analysed using EJ (eg Smith 2006).

Peña (2005:133) summarizes the relevance of applying EJ frames to park-associated ecotourism injustices well:

> Recall Guha's (1989) notion that the wilderness of the nature-appreciating ecotourist from the First World is the homeland of the displaced native in the Third World. The local is denied access to the means of right livelihood, the collective resources of the land, and the memories of place that sustain her identity, and all because of unjust acts of brutal enclosure for the sake of "economic development" or "wilderness preservation". Ecotourism is part of what Peña and others label a new postindustrial economy of so-called "non-extractive amenity industries". This new economy is embodied in a white spatial order that "cater(s) to a growing mass of 'aficionados' of cultural and natural diversity as objects of amusement, consumption, and appreciation" in a "global economy based on ski resorts, dude ranches, art galleries, museums and 'generalized elsewhere' associated with 'gentrification'". (Peña 2005:133)

Patterns of ecotourism development around the world are not random; ecotourism sites are systematically sited within certain types of places and communities (eg with attractive, "relatively unspoilt and natural" landscapes; with species of interest), many of which are marginalized communities of the Global South. Ecotourists, on the other hand, are mainly from the upper classes of the Global North (Brown et al 1997; Campbell 1999; Cater 2002; De Oliviera 2005; Grossberg, Treves and Naughton-Treves 2003; Hall 1994; McMinn and Cater 1998; Mowforth and Munt 1998; Pleumarom 1999; Weaver 1998).

Building on the work of Peña, Zebich-Knos's recent (2008) application of EJ to ecotourism in selected Latin American destinations comes closest to our application of EJ, in terms of topics and concerns raised. She uses EJ to contemplate whether "ecotourism businesses and national protected areas make good neighbors for local residents", questioning the fairness of ecotourism, an industry that in theory is supposed to be more directly beneficial for communities than mass tourism (Zebich-Knos 2008:280). She concludes that EJ can often "take a back seat" in top-down parks creation and related policymaking (Zebich-Knos 2008:186). Like us, she suggests that in practice, treatment of "the needs and expectations of communities adjacent to parks and protected areas ... proves to be a far cry from the theory" (Zebich-Knos 2008:187). Her analysis focuses mainly on the uneven distribution of ecotourism revenues and the limits to local participation in decision-making processes. She also questions the fairness of Mexico's Monarch Butterfly Special Reserve, where local residents are now inextricably linked to and dependent upon an international effort to save monarchs, with only limited seasonal potential for tourism revenue generation, while previous local activities such as logging are now illegal (Zebich-Knos 2008).

Although we acknowledge there is uneven distribution of ecotourism revenues in Tortuguero, we differ from Zebich-Knos (2008) in that we focus primarily on inequities associated with failed environmental impact management, as represented by the solid waste crisis. We raise questions similar to those forwarded by Tortuguero respondents about the potential primary costs (eg environmental costs; aesthetic damage) and secondary costs (eg economic costs; health costs; livelihood costs) related to ecotourism's negative environmental impacts on so-called "host" communities. Also, Tortuguero does not fit with Zebich-Knos's definition of ecotourism as being "mostly small scale" or mostly locally owned. Tortuguero received over 85,000 tourists in 2005 (De Haro and Tröeng 2006) and over 130,000 tourists in 2008 (Harrison 2009). Its ecotourism industry generates an estimated US$6.1 million a year (Troëng and Drews 2004), and none of the larger lodges in Tortuguero are locally owned (Campbell 2002a). Furthermore, although Zebich-Knos suggests that "Nowhere is the willingness to yield greater control to communities more evident than in nature tourism planning" (Zebich-Knos 2008:192), such willingness is elusive in Tortuguero.

Our Frame: A Focus on Marginalization
In this chapter, we isolate four characteristics of more conventional EJ communities and examine the extent to which these characteristics apply in Tortuguero, and other isolated ecotourism destinations. In this, our

chapter is somewhat distinct from others using EJ to explore new issues and new geographies. While most of these newer applications reflect the general issues at stake in conventional EJ, here we are also interested in the specifics, and the extent to which these are relevant in other scenarios. Despite EJ's emergence in a US-based context, we believe that its original concerns, albeit modified, remain relevant and offer a powerful lens through which to view our case. We are not suggesting that there is one, true, or static version of EJ; rather, we found our approach helpful for clarifying our own analysis and for allowing us to acknowledge the roots of EJ.

The modification we make is in our treatment of marginality. Rather than presenting a case in which race and/or class are the most obvious underlying causes for environmental discrimination and environmental injustices, we present Tortuguero as a less conventional type of marginalized community, one that can even be considered "rich" by some definitions. We therefore "push the limits" of EJ scholarship by applying an EJ frame centered on four characteristics related to marginality that can exist alongside more traditional race and class-related factors, and alongside relative wealth.

While waste problems are well trod ground for EJ, the application of EJ to ecotourism seems counter-intuitive at first. Ecotourism, in theory, protects and enhances local livelihoods and the environment (Boo 1990; Ceballos-Lascuráin 1996; Honey 1999). The maintenance of a healthy environment through ecotourism should benefit local residents as well as tourists, thus meeting expanded EJ goals (Walker and Bulkeley 2006). More recently, ecotourism has been redefined to include socio-political goals for "host communities" (especially indigenous communities, when relevant) such as community empowerment and increased local participation in decision-making processes (McDonald and Wearing 2003; Scheyvens 1999, 2002, The International Ecotourism Society no date; Zebich-Knos 2008). Ecotourism should therefore *prevent* environmental injustices through environmentally informed planning, proactive management, strengthened local livelihoods, and intimate local involvement in decision-making.

In spite of the seeming mismatch between ecotourism and EJ, we suggest that many communities experiencing ecotourism, particularly in the Global South, often share at least four characteristics with communities dealing with industries more commonly profiled in the EJ literature. We focus our analysis on these four common characteristics related to marginality:

1 The siting of the land use, industry, or facility is strategic; communities and places are chosen for specific geographic, political, and economic reasons;

2 The brunt of undesirable environmental impacts is borne by the community that hosts the land use, industry, or facility, and the community often gets little to no assistance from impact-causing parties to address such impacts. Conversely, investors and business owners, typically from outside of the community, tend to benefit most from the land use, industry, or facility;

3 The undesirable impacts of the land use, facility, or industry (and oftentimes the entire enterprise or components of it) are imported without well-defined community consent;

4 Affected communities often face political, economic, and social challenges to demanding compensation, reparations, or assistance *vis-à-vis* the siting of the land use, facility, or industry and its impacts.

Following a description of our case study, we describe how these four characteristics feature in more traditional EJ cases, how they are relevant to ecotourism in general, and how they play out in Tortuguero, Costa Rica. We discuss the extent to which EJ "works" as a frame in this case and what this might mean for EJ thinking more generally.

Case Study: Tortuguero, Costa Rica

The Village of Tortuguero

Tortuguero village is on the Caribbean coast of Costa Rica, close to the country's border with Nicaragua. Tortuguero's average daily temperature is 26°C (79°F) and average annual rainfall is more than 5000 mm (200+ inches) (Caribbean Conservation Corporation 2003). The village is located on a small strip of land between the Caribbean Sea and a lagoon and canal system, and draws its drinking water from groundwater close beneath the soil surface (Caribbean Conservation Corporation 2003). In 2004, 850 people lived in Tortuguero in 2004 (Jarquin and Gayle 2004), and the local population is now said to be over 1100 (not including an adjacent and growing squatter settlement called San Francisco) (personal communications with various respondents in 2008). Until the 1990s, however, the population was under 200 residents (Place 1991) and mostly composed of Afro-Caribbean immigrants from San Andrés island (Colombia) and Bluefields and the Corn Islands (Nicaragua) who settled from the 1920s onward. After several "boom and bust" economic cycles with corresponding population fluctuations, the village has grown (since the late 1980s) to include more recent and predominantly Latino immigrants from other areas of Costa Rica, Nicaragua, and Colombia, as well as North Americans and Europeans.

Tortuguero beach is the largest rookery for green sea turtles in the Caribbean and a US-based NGO, the Caribbean Conservation Corporation (CCC), has been conducting turtle research and monitoring there since the 1950s (Caribbean Conservation Corporation 2003). Turtle harvesting was once an important part of the local economy and culture (Lefever 1992; Place 1991), but commercial turtling ended when it was outlawed in Costa Rica in the late 1960s and when US and European markets closed in the 1970s (Place 1991), and very limited harvesting for community use ended in the mid-1990s (Campbell 2007). In addition to turtling, Tortuguero residents have, in the past, engaged in subsistence horticulture and fishing, and worked as laborers at nearby fruit plantations or in a short-lived lumber operation (Lefever 1992; Place 1991). Extractive activities were (and remain) severely limited with the creation (1975) and expansion of Tortuguero National Park (TNP), under the authority of MINAE (*Ministerio del Ambiente y Energia*), the national ministry of environment and energy (Campbell 2002b, 2003; Evans 1999; Troëng and Rankin 2005). In 2002, amendments to the Costa Rican law protecting turtles (Law 8325) made all turtle-taking and trading offenses punishable by jail, in addition to fines (Campbell 2007; Goodier 2005; Silman 2002).

TNP sits adjacent to the southern end of Tortuguero village, comprises 26,156 hectares of land and 50,160 hectares of marine territory, and protects 22 miles of turtle nesting beach (Caribbean Conservation Corporation 2007; MINAE no date). While the creation of TNP has limited local access to resources, it also plays a key role in fueling the latest version of Tortuguero's economy: ecotourism. TNP's environment (eg lush forest; lagoon and canal system) and wildlife, especially nesting sea turtles, are the biggest tourism draws (Lee and Snepenger 1992; Meletis 2007). Ecotourism to Tortuguero began in the late 1980s, with intrepid travelers such as visiting biologists and other nature-oriented explorers (Lee and Snepenger 1992; Lefever 1992; Place 1991); it now caters to a range of ecotourists who either stay in lodges (higher-end accommodations located outside of/adjacent to the village proper) or *cabinas* (lower end, more basic accommodations in the village). Whereas only a few hundred people visited TNP annually in the early 1990s (Place 1991), it is now visited by over 130,000 tourists a year (Harrison 2009), including more mainstream "package ecotourists".

For all intents and purposes, Tortuguero's economy can be considered 100% ecotourism based. Existing businesses either serve tourists (eg lodges, *cabinas*, restaurants) or employees of the tourism industry (eg grocery stores catering to local residents), and tourism provides jobs and entrepreneurial opportunities. Turtle tour guides who accompany tourists to the nesting beach at night can make US$100–200 a night (Peskin 2002). Troëng and Drews (2004) estimate that Tortuguero's ecotourism industry generates US$6.1 million gross revenue per year, and there are no other sizeable industries in Tortuguero. The advent and growth of ecotourism in Tortuguero has contributed to improved infrastructure and improved connectedness with surrounding areas through, for example, increased and improved transportation of people and goods in and out of the village. Tourism revenue has also supported community projects and infrastructure improvements. Guide money and other tourism-derived funding helped to create an aqueduct water system for the village, a playground, and a combined sports hall/assembly structure. The village's regional prominence, its relative wealth and its growing population, which are all related to tourism, have also facilitated and partially funded an on-site clinic building with a (shared municipal) staff, operating twice-weekly.[4]

Despite a reputation for successful turtle conservation and ecotourism, Tortuguero is also experiencing associated environmental problems. While there have been no formal environmental assessments in Tortuguero (with the exception of studies relating to sea turtles, birds, and other wildlife, occasional testing of the local drinking water, and a waste audit that will be described in detail), environmental issues are visible on the land or water, and are of concern to residents (Meletis 2007). These include bank erosion, canal siltation, and other water quality issues (associated with increased boat traffic, much of which is related to tourism); motor-boat related environmental concerns (habitat disturbance; noise pollution; water pollution); declining sightings of certain wildlife species (eg manatees); legal and illegal deforestation for construction; illegal harvesting and construction (eg the squatter settlement) within the Park; sewage-related problems (eg lack of sewage treatment; direct pumping of sewage into the lagoon; running out of space for septic tanks); and solid waste management shortcomings and resulting environmental impacts (and potential health impacts) (Meletis 2007). Such challenges are often compounded by the fact that there is no elected on-site political authority in the village. Despite its increasing size; Tortuguero remains part of the regional municipality of *Pococí*, headquartered in the city of *Guapiles* (several hours away). The Park administrator's office is also located in *Guapiles*.

The History of the Waste Crisis
Before ecotourism became the dominant industry in Tortuguero, the village and its population were smaller (200 people in the early 1990s; 850 people in 2004; over 1100 people in 2007) (Jarquin and Gayle 2004; Place 1991). From its founding in the 1930s until the early 1990s,

village residents produced most of their own food and the cash economy was limited (Lefever 1992). Local waste disposal practices of burying or burning wastes in residential yards or common areas were probably appropriate for the amounts and types of wastes being generated. With fewer connections to the external cash economy, there were fewer packaged goods, fewer plastics and other synthetic or non-biodegradable materials, and fewer people generating solid waste. Burying and burning are no longer feasible given: the increased volume of waste generated by over 800 residents and over 80,000 tourists a year when this research was conducted in 2004–2005 (Jarquin and Gayle 2004); and changes in waste stream composition to include greater percentages of non-biodegradable materials such as plastics, and more hazardous or otherwise problematic wastes such as disposable batteries.

Local waste management has been especially contentious and challenging since the inception and growth of ecotourism, since both the size of the village (population size; built environment; village density) and the volumes of solid wastes generated have increased along with the growth of ecotourism. During the waste crisis observed in 2002–2004, Tortuguero's "recycling plant" (described in an upcoming section) was often closed because of lack of funding; lack of staff; lack of local compliance; conflicts over the running of the plant; incinerator-related issues, and so on. During closures, residents and businesses returned to traditional methods of waste disposal, including burying, dumping, and burning waste in yards, on the beach, and in other public areas (Camacho 2003; personal observation 2002–2004). Some residents, businesses, and lodges continued to leave solid wastes at the plant when it was clearly closed. The waste crisis was reflected on the landscape (Figures 1 and 2), and visible to residents and tourists.

Before 2000, no organized waste disposal system or plan was in place in Tortuguero. In 2000, a "mixed commission" was formed to deal with waste management, with the participation of various national authorities and the municipality of *Pococí*. As a result of this commission, and with assistance from JAPDEVA (a regional government-funded coastal development agency) and European Union-based funding, the *Planta de Tratamiento Integral de Desechos Solidos* (the integrated solid waste treatment plant) was constructed in 2000 (see Figure 3). The plant is located on one of the main footpaths in the center of the village, near houses, stores, a playground, and other common areas. Locally called the plant (*la planta*), it was originally to be run by the Women's Association in the hopes of improving local waste management and providing women with new skills and an enterprise of their own.

From its first days, the plant had problems (Camacho 2003). The plant workers' main tasks were to separate wastes, and then segregate

Figure 1:   A tourist looking at a dump site adjacent to the beach in Tortuguero (source: photo by Zoë A Meletis)

Figure 2:   "Open burning" of waste in the village in 2004 (source: photo by Zoë A Meletis)

Figure 3: The plant in 2000 (source: photo by J. Peskin)

and prepare valuable recyclables (eg aluminum cans) for transportation out of the village. Transportation quickly became problematic; the plant had no boat of its own and the vendors who bought and transported out recyclables did so unreliably, creating storage issues for the plant, and potential environmental and health risks associated with storing large amounts of waste outside, in a central location. There was no official plan or network to integrate efforts in Tortuguero with greater municipal waste management services and infrastructure (Camacho 2003).

The plant's original design also included a "biodigester", which employees claim never worked, and was mainly used for storing waste in 2003 and 2004. Materials deemed non-recyclable, including plastics and baby diapers, were burned in the plant's incinerator. This incinerator was not equipped with necessary or functional devices for filtering its exhaust, and was likely emitting dioxins and other undesirable and toxic compounds into the air (Camacho 2003). The plant's incinerator was still in use in 2002–2004, but was experiencing increased slowdowns and breakdowns. It was rusted, with its stack leaning and billowing dark and malodorous smoke that left ashy deposits around town. Residents voiced concerns about the incinerator's possible negative impacts on air quality and health of the villagers (eg increased asthma rates) and plant workers complained that, at best, the incinerator could burn for a few hours before malfunctioning. Plant workers worked without masks and

Figure 4: The plant in summer 2004; it was largely reduced to storage (source: photo by Zoë A Meletis)

gloves, and were exposed to unknown health impacts by their proximity to a malfunctioning incinerator.

By 2004, the plant and its workers faced increasing numbers of operational issues including changes in management and related conflict, non-payment for services, an inability to pay plant workers, plant equipment thefts, and uncooperative plant users (eg people refusing to separate solid wastes into types). As a result, by the end of the summer of 2004 the plant served for little more than storage (Figure 4). Only one man continued to work at the plant, and he was limited to collecting and bagging wastes (Meletis 2004). Concerns about potential health impacts drove the director of the health clinic to make an official complaint to the municipal tribunal in 2004. He argued that, under Costa Rican law, the municipality shares the responsibility to provide a "healthy environment" for its citizens and was therefore neglecting its responsibilities.

Public concern regarding the waste crisis eventually led to a government-sponsored audit of the plant (referred to in our chapter as Camacho 2003). Despite some methodological shortcomings,[5] it offers some insights into waste generation in Tortuguero. For example, even though only three of the more than 15 lodges were sending waste to the plant at the time of the audit, lodge-generated waste represented more than half of the waste received daily. The three participating lodges disposed of approximately five times the amount of plastic bottles and more than twice the number of plastic bags as did the community. Lodges also generated more "other plastics", newspaper, cardboard, and glass. In general, the lodge waste included more recyclables and more non-biodegradable wastes, while the community deposited a greater amount of organic waste (Camacho 2003). This audit only captured a fraction of the tourism industry's role in the solid waste crisis since few lodges and their wastes were accounted for, and given the ratio between the numbers of visitors to local residents. Based on these results, Camacho (2003) suggests that the number of tourists per year may represent a threat to the balance of the local ecosystem, as reflected in the village's inability to properly manage tourism-associated solid waste. He lists the following types of shortcomings in Tortuguero related to waste management: cultural/social (eg no ongoing waste-related education programs; apathy towards the waste problem), institutional (eg inadequate municipal support), financial (eg a lack of resources and the lack of authority to enforce payments), and technical (eg deficiencies in existing services). Camacho insists that the clear result is a "need for immediate action on the part of actors involved" to solve this problem (Camacho 2003:20).[6]

Framing Tortuguero's Solid Waste Crisis Using Four Characteristics of Conventional Environmental Justice Communities

During Meletis's research, a popular opinion among respondents in Tortuguero was that the distribution of environmental costs and benefits associated with ecotourism is unfair, and that powerful tourism actors like the lodges are responsible for creating and ignoring environmental problems. One respondent answered a question about whether or not the environment was being well managed as follows: "Completely not, no. Why? Because it's uh... Like with other parts of the world—the big fish eats the small one; the rich one controls the poor one; the one with more money has more rights and that's the way it is in Tortuguero" (R52 2004). To what extent does Tortuguero's solid waste management crisis represent an environmental injustice and to what extent is an EJ frame useful for understanding the crisis? In this section, we frame the Tortuguero case using four characteristics shared by Tortuguero and more conventional EJ communities:

1. The siting of the land use, industry, or facility is strategic; communities and places are chosen for specific geographic, political and economic reasons.

In more conventional US-based EJ sitings of heavy industry or noxious facilities, isolation and distance play two roles. Due to NIMBYism (Not In My Backyard syndrome), communities and their representatives resist undesired planned developments by trying to push unwanted industries and facilities into other places. Many LULUs (Locally Unwanted Land Uses) and their impacts are pushed into marginal (mostly visible minority) neighborhoods, away from affluent (mostly white) neighborhoods (Bullard 1990). As a result, negative impacts are mainly isolated in communities distanced from power and decision-making. In ecotourism settings, on the other hand, isolation, remoteness, and distance from urban centers are part of the appeal of ecotourism destinations, and contribute to the appearance of untouched nature (Mowforth and Munt 1998; Nelson 2005; Wight 1996). The isolated, "seemingly undeveloped" image is often reinforced by inadequate infrastructure or low levels of "modernization" in terms of built form (Akama 1996; Carrier and Macleod 2005). Perhaps ironically, ecotourism's impacts and a community's inability to deal with them are often aggravated by ecotourism's frequently remote locations and related infrastructural shortcomings, the very same geography that acts as a draw and preserves the environment "enough" for ecotourism to take root in the first place (Hillery et al 2001; Meletis and Campbell 2007).

Geographic isolation is central to Tortuguero's image as an ecotourism destination; with no access by road and no cars in town, Tortuguero lacks a central modern icon—the car—and thus seems more "natural". Lack of easy transit to Tortuguero's also satisfies ecotourist desires to feel "off the beaten path". Limited transportation also complicates waste management. First, it is difficult to move waste out of Tortuguero. Second, it is difficult to access outside support, networks and resources that might contribute to managing impacts, such as municipal services and markets for recyclables (Camacho 2003). Finally, its lack of roads has been used by the municipality of *Pococí* as an excuse for failing to provide waste collection and disposal services; it claims that its responsibilities end where road access does. Tortuguero's isolation contributes to the neglect by the regional municipality, in terms of keeping its impacts and struggles "out of sight".

Waste management presents a "landscape challenge" for ecotourism destinations, as communities must protect the "natural" image associated with their isolation. Thus, although ecotourism generates waste in Tortuguero, it must appear not to (Colten and Dilsaver 2005; Thrupp 1990); ecotourists do not expect to look upon waste or related infrastructure (Carrier and Macleod 2005; Hughes and Morrison-Saunders 2003; Ryan, Hughes and Chirgwin 2000; Urry 1995, 2002). Hiding waste and its management in Tortuguero is challenging, however, because: the area includes a "real" village where people live, work, and generate waste; while most tourists stay in lodges outside the village, tourists come through the village proper (eg on village tours) and can potentially wander around most parts of the village; and Tortuguero is located on a narrow land strip between two bodies of water, with limited capacity to "hide" waste and waste-related infrastructure "backstage". Thus, ecotourism places a double burden on the host community: a set of impacts that are difficult to manage and that must remain hidden, and a location where the same elements that make it attractive as an ecotourism destination limit local abilities to manage impacts. These burdens are similar to those often seen in more traditional EJ scenarios in which communities were targeted as sites for "development" because of their location, their resources (eg cheaper land), and community composition, and as a result, they have greater difficulty fighting against a use or its undesirable impacts.

In traditional EJ cases, economic marginality is often a key theme, with LULU-related development or facility siting offering economic options where there are few others. LULUs are "sold" to communities in the face of no other employment and income options (Lee 1993). Establishment of a LULU, however, can further limit options. In the case of Swan Hills, Alberta, for example, the community voted in favor of accepting a hazardous waste facility in the face of failing traditional industries. Much needed revenue and employment was enough to convince the majority of the community that it could live with a LULU. However, once the hazardous waste facility was in place, development prospects were further limited by Swan Hills' new place image. "Clean" industries did not want to develop or invest near such facilities. In subsequent years, new developments were limited to additional (similarly) hazardous industries that fit Swan Hills' development profile (Bradshaw 2003).

In the case of ecotourism, the same geographic isolation adds to locations' attractiveness as destinations are often linked to their economic development histories (or the lack thereof). Ecotourism is seen as a "replacement economy" for industries that have preceded it. Natural resource extraction and/or processing may have been limited in the past because of the distance to or lack of accessibility by industrial centers. Or extractive activities that were in place may have declined (Campbell, Gray and Meltis 2008; Nygren 2003; Thrupp 1990). The resulting quality of the natural environment today makes these areas targets of conservation programs, and ecotourism is one of the few economic development activities widely deemed compatible with conservation (Meletis and Campbell 2007). Therefore, ecotourism is often part of a *very limited* set of development options (Mowforth and Munt 1998; Weaver 2001). Part of the "limited options" scenario, as is the case in more conventional EJ cases, ecotourism is often sold on employment and revenue (Troëng and Drews 2004) and may be accompanied by efforts to limit any remaining resource extraction. Once in place, ecotourism can impose a restricted development profile on communities because its aesthetic requirements may exclude industries that compromise the ecotourism landscape. Ecotourism-based communities are therefore limited to types of development that, at least aesthetically, appear to be "natural" (Akama 1996; Carrier and Macleod 2005). These types of limitations are evident in Tortuguero. For example, debates about the utility of extending a road to the village weigh the benefits for the community through increased access to services against damage to the ecotourism aesthetic based on a carless, remote location.

2. The brunt of the associated undesirable environmental impacts is borne by the community that hosts the land use, industry, or facility, and the community often gets little to no assistance from impact-causing parties to address the impacts. Conversely, investors and business owners, typically from outside of the community, tend to benefit most from the land use, industry, or facility.

In many traditional EJ cases, outsiders such as extra-locally owned companies or various levels of government introduce LULUs into communities, and reap most of the profits. Despite initial promises of revenue generation, employment creation and "trickle-down" multiplier effects, the proportion of jobs and multiplier effects that remain in host neighborhoods typically represents a fraction of overall profit. In contrast, people working and/or living nearby tend to bear the direct environmental impacts and health costs, as well as associated financial costs. Bullard (1990) profiles many such cases to illustrate the nature of repeated patterns of unjust siting practices that exist within the USA, and the commonalities between EJ cases and the types of affected communities.

The unequal distribution of costs and benefits is also evident in ecotourism. Many prospective or current ecotourism-based communities are in places that outside business interests (eg foreign investors) seek to develop, that conservation actors (often international or foreign NGOs) seek to preserve, and that offer an attractive ecotourism landscape for foreign visitors. Ecotourism therefore emerges as a potential way to meet conservation goals, related to location and natural resources, while also satisfying some local development goals, related to the state of the local economy (Campbell 1999; Gray 2002). Despite progressive definitions of ecotourism that often include a focus on local ownership or empowerment, ecotourism development often reproduces North–South and class-based power imbalances, and associated uneven distributions of costs and benefits (Campbell, Gray and Meletis 2008; Weaver 1998) in ways parallel to more conventional forms of "development" in traditional EJ cases. Furthermore, many of the critiques of ecotourism as an elite tourist activity with limited access for the wealthy few (eg Akama 1996; Cater, 2006; Mowforth and Munt 1998; Thrupp 1990) parallel EJ critiques of mainstream environmentalism as exclusionary (Bullard 1993; Pulido 1996).

In general, tourism includes high levels of leakage (loss of revenues to "the outside"), and ecotourism is no exception, often as a result of high levels of foreign or extra-local ownership (Bhattarai, Conway and Shrestha 2005; Scheyvens 2002; Weaver 1998). In Tortuguero, despite estimates of the considerable revenues generated (see eg Troëng and Drews 2004), there is evidence that high levels of leakage are also occurring (Campbell 2002a). All of the lodges, which house the majority of tourists to Tortuguero, and several successful businesses are extra-locally owned, many tourism-related purchases (eg plane tickets; tour or weekend package tickets) are purchased outside of the village, and there are significant barriers to entry for local investors (Barrera 2003; Campbell 2002a). Furthermore, there is no local bank on site to capture deposits or to facilitate local investments. Finally, many of the local ecotourism jobs are relatively menial positions and there are limits to upward mobility (Barrera 2003).

While leakage means that tourism profits largely accrue outside of Tortuguero, most of the solid waste generated by the lodges, the *cabinas*, the small businesses and the tourists themselves, stays behind. Given that the lodges house the majority of the 130,000+ tourists (Harrison 2009) who visit Tortuguero (75%, according to Lee and Snepenger 1992), it is not surprising that they and their guests generate the bulk of the volume of waste (Camacho 2003). While some lodge managers undertake efforts to reduce wastes, plant workers suggested that lodge participation in plant-led efforts was variable, with some lodges refusing to use plant services, refusing to pay, or dumping waste at the plant without the correct payments. Some residents are concerned that lodges are engaging in illegal dumping on or near lodge land—a practice that creates potential environmental and health impacts for the greater Tortuguero environment.

Tourism-generated waste can be considered "imported waste", a term used in other EJ cases (Fletcher 2003; Wu and Wang 2002), in three ways. First, tourists on vacation generate waste in their travel destination rather than at home, and these destinations are often less equipped to deal with waste and related impacts than tourists' home communities. Second, some waste generated by tourists is imported from their homes (eg batteries and film cases). Third, waste is imported into the village from lodges outside of the village when lodge tourists dispose of some wastes while *in* the village, and when lodges send their wastes to the plant. In all three cases, disposal and related costs occur *in* the village, while the original purchases (and revenues) do not—the costs and benefits are geographically separated and accrue to different actors.

Furthermore, as in some other EJ scenarios, ecotourism and its supporting infrastructure offers limited direct-use benefits for Tortuguero residents or even Costa Rican nationals. As ecotourism has grown, the percentage of domestic visitors to TNP has decreased, from 56% in 1986 to only 12% in 2004 (Harrison et al 2004; Place 1991). Few residents use TNP facilities for their own enjoyment,[7] and they have lost access to previously used resources. During sea turtle nesting season (May–November), nighttime beach access is limited to permitted turtle guides and paying tourists; local access is prohibited. However, residents have benefited from various projects that included tourism-derived or tourism-directed funds. These include improved private transportation infrastructure and services, though service providers sometimes prefer to carry tourists, who pay higher rates. Residents have also petitioned for and successfully obtained state-funded communication improvements as a result of local tourism success and related commercial and residential communication needs.[8]

3. The impacts of the land use, facility, or industry (and oftentimes the entire enterprise or components of it) are imported without well-defined community consent.

In traditional EJ cases, problems arise *vis-à-vis* community consent to land uses within their neighborhoods or surrounding areas. As discussed earlier, notions of "choice" are constrained when there are no other economic options (Bullard 1990), but confusion can also exist about the kinds of development that people are consenting to. For example, Northwood Manor residents in Texas thought that they were getting a shopping center or new homes for their subdivision when they were actually getting a landfill (Bullard 1990). Sometimes what has been consented to only becomes apparent once the facility is in place and its "impact shadow" and associated costs begin to emerge; some impacts take years or decades to become apparent. Therefore, even if communities do initially consent, they are unlikely to consent to *all* of the impacts associated with a facility since some impacts, secondary impacts, and associated costs only emerge after the fact.

Questions of consent are relevant in Tortuguero. TNP, for example, was decreed from the capital (San José, Costa Rica) without the direct consent of local people but with support and pressure from the capital, the USA (Evans 1999), and beyond. Some resentment lingers over this lack of proper community consultation and the lack of adequate compensation for people whose land was expropriated to create TNP.

One respondent expressed his frustrations with current Park–community relations this way:

> The park system is managed by the government and they don't care about what we say . . . The park is an imposition for us. It's good that they have their laws but they shouldn't affect us, we are not visitors, we are from here . . . This problem of the park that we are talking about has created resentment; there is a division that is like a divorce between the park and the people. There is enmity between us and the conservation. Because why? Because they apply the same laws to us as to the visitors (R72 2004).

Ecotourism is currently tied to TNP, but where ecotourism differs most from traditional EJ industries is that it often "just happens", with little coordinated planning or foresight, unlike the siting of heavy facilities or industries that often require permitting, planning and documentation. The first lodges in Tortuguero were established by foreigners to serve the few hundred tourists who visited before the tourism boom. Through a variety of processes over which residents in Tortuguero had little control (eg the global interest in and increased growth of ecotourism, Costa Rica's national development strategy, the popularity of turtle viewing with ecotourists; see Campbell 2002b), ecotourism to Tortuguero increased dramatically with little centralized effort and without a local tourism development plan.

While residents appreciate tourism and its benefits, they are concerned about the lack of control over the industry and its associated impacts, and the lack of genuine opportunities to participate in decision-making processes regarding ecotourism and environmental management (Barrera 2003; Meletis 2007). Many local respondents reject the current state-of-affairs of the environmental impacts of ecotourism and their management. Many cite the lodges as being largely responsible for environmental problems such as the waste crisis, and for neglecting to do much about such problems. In response to questions about the benefits and threats associated with tourism, one man said the following: "The hotels. They fill up the most, [they have the money], but in the village, there isn't money, the only thing that comes to the village is garbage, nothing more . . . The [hotels] should help the people in the community. The threat is that they don't help" (R59 2004).

Residents did not consent to unmanaged environmental impacts of ecotourism, or to paying financial costs associated with unmanaged environmental impacts of ecotourism. They see waste-related impacts *in* the village as primarily being the responsibility of those actors that are benefiting greatly from tourism such as the lodges and the Park, and they reject the use of the village as a "sacrifice zone" for the industry's impacts. As one woman said: "If we don't look after the natural resources here, very soon there won't be tourism" (R74 2004). While the term sacrifice zone is used to describe areas in tourism destinations in which visitation and its associated impacts are purposefully located (Weaver 2001), EJ scholars use the same term for areas and communities that are faced with more than their fair share of environmental costs (Bullard 2001; Bullard and Johnson 2000). Tortuguero, according to local respondents, represents both meanings of the term.

Ecotourism in Tortuguero continues to develop without a plan and without formal consent by local people. The political distance, both figurative and literal, between Tortuguero, municipal authorities, and the national capital contributes to local perceptions of municipal and national government neglect of village concerns. Several respondents voiced feelings of powerlessness in terms of their ability to change the course of ecotourism development in Tortuguero or influence lodge-related development (Meletis 2007). Because residents are mostly relegated to lower level jobs within the industry (Barrera 2003), their abilities to change or guide ecotourism development from within are also limited. Explaining the lack of individual power and suggesting that solid waste management could be improved through united pressure on the lodges, one man said this: "The second thing that I would do is pressure the big hotels to help the village. But I can't do anything, but that's what I would do. If I called people and said "let's call the hotels and pressure them", they would say, "no, no, that's ok'...But that is what we should do" (R52 2004).

4. Affected communities often face political, economic and social challenges to demanding compensation, reparations, or assistance vis-à-vis *the siting of the land use, facility, or industry and its impacts.*

EJ academics and activists draw attention to the political, economic, and social challenges that marginalized communities face in preventing environmental injustices (Hamilton 1993), which position them as "powerless to resist" (Chavis 1993) and include obstacles to having their voices heard after a LULU is in place. Local testimonials and evidence are often discredited as being anecdotal, and residents are contrasted with "scientific experts" (Edwards 2002; Simpson 2002). Čapek (1993), for example, documented the struggles waged by the community of Carver Terrace in Texarcana, Texas, for "the right to information", "the right to a hearing" and for "democratic participation". Within the Carver Terrace community, marginality made community members reliant on outside and sometimes dubious "experts". Often, they did not receive critical information in a timely and direct manner, they also had difficulties presenting evidence, had community-based evidence dismissed by various authorities, and lacked formal opportunities for input (Čapek 1993).

Ecotourism is a labor-intensive industry, and labor demands can constrain the community's ability to organize for improved environmental management, or to resist further development. In Tortuguero, some people work at lodges during the day and as turtle guides at night, leaving them few free hours. Occupations such as guiding also demand that workers be available during certain times everyday, limiting their ability to organize or leave the village. Small business owners in Tortuguero stand to lose considerable amounts of income if they take time off or leave the village for a day or two, especially during high season. The highest tourist seasons, which are the busiest work periods for many residents, also represent peak solid waste generation times. Therefore, conflicts between individual work schedules, tourist season constraints, and daily responsibilities limit many local peoples' abilities to contribute to waste management meetings or other activities.

Challenges to organizing or uprising in an ecotourism community go beyond mere labor-related constraints and political challenges. As is the case in Tortuguero, local access to various forms of capital and training is often limited and members of communities cannot easily establish or fund local businesses (Barrera 2003; Place 1991). They are often economically and politically vulnerable to control by "outsiders" and thus less able to resist, reject, or challenge certain development components or undesirable impacts. Residents are also often conceptualized as "part of the ecotourism product" rather than simply as labor for an industry. It is difficult to act up, resist, and make demands regarding local environmental justice while preserving an idyllic ecotourism image. Taking public action might compromise the destination image and, ultimately, local livelihoods (Carrier and Macleod 2005; Levi 1999).

Discussion
The analysis presented here illustrates the compatibility, appropriateness, and utility of EJ as a theoretical frame for understanding the disproportionate burden that the solid waste management problems in Tortuguero, and negative ecotourism-associated impacts in the Global South in general can represent. As with traditional EJ cases, the siting of ecotourism as an industry is strategic and important, there are inequities in the distributions of costs and benefits, the industry and its impacts are imported with adequate community consent, and the community's ability to address the challenges associated with negative impacts is constrained by factors associated with its marginality. Our analysis illustrates that there is variation in how these characteristics play out (eg traditional EJ case studies typically confront industries arising in already "degraded" environments, while our case study concerns an

industry that relies on "pristine" ones), but the fundamentals are the same. In this section, we turn our attention to the shortcomings of the EJ framework, as we propose it. More specifically, we consider five characteristics of the Tortuguero case that are not easily accounted for using a conventional EJ frame. With this in mind, we suggest ways in which EJ might be expanded to accommodate these.

Widespread Benefits, Despite Leakage

Unlike industries in traditional EJ cases, most Tortuguero residents support the presence of ecotourism and appreciate its benefits. A respondent who works as a turtle guide, among other jobs, explained tourism's benefits in this way:

> With respect to tourism, it is really important for the village in order to lift the village up [its economy], it needs more revenues and it is a well welcomed tourism. It is also growing, every year more people are coming to see Tortuguero's turtles and nature. (R58 2004)

Benefits are mostly employment related, but also include entrepreneurial opportunities and infrastructure improvements. Despite resident concerns, most residents would be unlikely to choose a return to life without ecotourism.

Widespread benefits and appreciation for the industry, above all else, is typically lacking in most conventional EJ scenarios. Conventional LULUs are often associated with a smaller proportion of jobs for the community, and more concentrated clusters of local benefits (Bullard 1993; Maschesky 2003). It is also unusual for conventional EJ industries to provide the entire underlying economic base for a community, as ecotourism does in Tortuguero.[9]

Marginality in the Context of Relative Wealth

Like most traditional EJ communities, Tortuguero is marginalized in many ways. However, while its economic options are limited, ecotourism has brought wealth to the community and Tortuguero is considered to be relatively well-off compared with its neighbors. Tortuguero is the most successful rural tourism destination in the province of Limón and it has greater employment and revenue generation opportunities than any of the surrounding villages. This wealth does not appear to reduce Tortuguero's marginality when it comes to seeking support in dealing with waste management challenges, however. Many residents believe that Tortuguero's relative wealth in the region fosters neglect from the municipal authority. While conventional EJ communities are neglected on a variety of levels for various reasons, their relative regional wealth is not typically one of them. Thus,

Tortuguero's relative wealth is not easily accommodated in an EJ frame. However, relative wealth and the presence of a "successful" industry do not guarantee an equitable distribution of costs and benefits or preclude the production of environmental injustices. Nor do they overcome the issues associated with geographic isolation and the community's distance from sources of potential support. If local respondents are correct in their belief that it is because of their relative wealth that they are neglected, then it is not just a case of environmental injustices occurring in *spite* of wealth, but *because* of it. This scenario, where a marginalized community and its problems are neglected *because* of its relative wealth within an otherwise marginal and impoverished area, is new for environmental justice.

A Lack of Unified Community Opposition

Environmental justice often relies on an "us versus them" frame, in which a seemingly united community fights against an environmental "bad guy" responsible for undesirable impacts. In our case, while there is some discursive use of "us versus them", the story is more complicated. The local politics and infighting that complicate the waste crisis in Tortuguero (Meletis 2007) cannot adequately be captured in an EJ frame. The perceived lack of village cooperation was highlighted by an older respondent when discussing the state of the beach:

> Well, the beach has turtles but it is very dirty because no one wants to cooperate to clean it up. So right now, the beach is dirty and the village too, because the people don't want to cooperate in anything—the kids don't, no one does, not even the teachers. Because if the teachers said "let's go clean up the beach", the kids would do it. (R67 2004)

Despite common concerns about the waste crisis, and the shared goal of improved waste management, the community is divided over how it should be done, who should do it, and who has the authority to initiate changes. Various political camps supporting different informal leaders exist in the village. This mosaic of opinions is not easily incorporated into a traditional EJ frame because the latter tends to emphasize the community's unity in opposing the industry or facility at hand. This may stem from the fact the EJ literature often focuses on organized activism within a community, rather than the community as a whole.[10]

The diversity of opinions in Tortuguero is an important part of the waste crisis in that it complicates waste management efforts and acts as an obstacle to village-level cooperation with respect to any new waste management attempts. Conceptual space must therefore be created within the EJ frame to accommodate internal differences within communities in understanding local environmental problems and responses to them. Without room to consider how local "crab antics" in the village (Lefever 1992) complicate waste management in Tortuguero, it is difficult to understand the nature of local relationships, and to identify possible alliances and conflicts with respect to waste management.

"Us versus Us"

The solid waste crisis in Tortuguero includes an important element of self-blame among residents, who see themselves as contributing to the solid waste crisis alongside other actors such as the lodges and the Park. They use phrases such as "*no tenemos culture*" (we don't have culture) or "*faltamos educación*" (we lack education) to explain their lack of waste-related and general environmental awareness (Meletis 2007). A second element of self-blame relates to the community's failure to unite in its efforts to address waste management. Respondents pointed out problems with cooperation and compliance within the community, and gossiped about who does or does not contribute to local waste management efforts. For example, some respondents were highly critical of past efforts to run the plant (Meletis 2007). Thus, while local respondents laid the majority of blame for waste management problems at the feet of the government and the tourism lodges, they also recognized their own failure to contribute. One respondent explained it this way:

> Yes, well, what happens is some people help and others don't. So we don't need more people here, we just need to [fix our problems] amongst ourselves. I don't think it's good, for example, that people like students from the US come here to see nature and see dirty streets and garbage. [The problem changes every year here] and we need to [understand] the problem and how much garbage there is. (R73 2004)

Another woman explained how local passivity might play a role in ongoing waste management challenges: "I think that it's ourselves. We are not organizing ourselves, we are just remaining calm. There is no respect" (R57 2004). This self-blame does not yet fit well within an EJ frame despite the recognition by some scholars that communities members are involved in shaping EJ scenarios, rather than simply being passive victims of them (eg Harvey 1996; Njeru 2006; Pellow 2000).

The Consumer as Overlooked Fourth Party

In traditional EJ cases, three actor groups are typically highlighted: the community and key activists within it, the government, and the industry, company, or facility. An EJ frame does not typically extend to the consumer of the product or service that is causing environmental injustices, as consumers are often geographically removed from the site.

This is arguably a general weakness of EJ, but it becomes particularly problematic in the case of tourism, where consumers come to the site of production/consumption and their roles in the problem at hand are visible and direct. Tourists are clearly implicated in Tortuguero's waste problems, but local residents themselves tend to overlook this. Respondents generally described tourists as being different, ie gentler, more eco-friendly types of tourists and no respondent directly associated the waste crisis with the behavior of visiting tourists as a group (Meletis 2007). Perhaps tourists escape blame because they are not associated with tourism or waste management. Or, perhaps tourist–resident interactions are generally positive and do not elicit feelings of distrust or resentment similar to those that some residents harbor towards government and the lodges. Another possibility is that residents do not openly blame tourists for fear of losing tourist dollars. Further research is needed to tease out explanations for the lack of voiced connections made between tourists and the waste crisis. In Tortuguero, tourists also fail to implicate themselves in waste problems (Meletis 2007).

Incorporating the tourist into an EJ framework may be a possibility on the horizon. EJ scholars have recently begun to extend investigations of environmental injustices to include the consumer (eg Martinez-Alier, 2003; Rees and Westra 2003). Work on the political ecology of consumption also offers promise for extending the analysis of tourism-associated impacts to tourists. Recent works conceptualize tourists and other consumers: as individual and collective impact producers and therefore partly responsible for the impacts of the industries that cater to them; and as political agents, with the (potential) power to change individual and collective contributions on a larger scale (Bryant and Goodman 2004; Goodman 2004; Heyman 2005; Meletis and Campbell 2007; Njeru 2006). It is also part of an important debate over whether alternative modes of production and consumption, such as ecotourism and the fair trade industry, are changing social and environmental relations, or simply reinforcing existing inequitable relationships, with little real change (Bryant and Goodman 2004; Campbell, Gray and Meltis 2008; Charnley 2005; Duffy 2002; Goodman 2004; Lyon 2006; Manokha 2004; Moberg 2005; Mowforth and Munt 1998; Talbot 2004; Tarmann 2002). As we have illustrated here, EJ offers a useful frame for positioning ecotourism in this larger debate.

Conclusions

This chapter offers an unconventional use of EJ by focusing on an ecotourism case study, an industry not typically associated with environmental injustices. We provide an example of how tourism scholars can adopt concepts from other fields and inject theory into tourism cases studies in order to help illuminate key issues, and to contribute towards increasing the theoretical rigor of the field. This unconventional use of an EJ frame to contemplate ecotourism and the solid waste crisis in Tortuguero makes several contributions to our understanding of EJ, of ecotourism, and of Tortuguero.

First, we illustrate the appropriateness of applying EJ to ecotourism. Although they may appear different on first glance, ecotourism-based communities in the Global South share characteristics with conventional EJ communities. For example, while both types of communities may be slated for "development" because they are marginalized, traditional EJ cases often take place in hyper-urban environments considered degraded or less desirable. Ecotourism, on the other hand, is situated in seemingly pristine environments. In the Global South, however, such environments are often inhabited by communities facing constraints, challenges, discrimination, and limited sets of development options similar to those faced by conventional EJ communities.

Second, we illustrate some of the new issues that can arise when EJ is expanded. For example, we highlight the self-blame evident among residents, and how this is not easily accommodated in a traditional EJ approach. We suggest that one way ecotourism differs greatly from traditional EJ industries is that ecotourism involves local community members figuratively selling themselves as part of the village aesthetic (Ateljevic and Doorne 2005; Mowforth and Munt 1998; Nelson 2005; Smith and Duffy 2003). They are often depicted as stewards of their environment and romanticized as being "in tune" with nature (Bryant and Goodman 2004; Norton 1996). In Tortuguero, local people are proud of their environment in spite of their environmental management concerns (Meletis 2007). They are also aware, however, that a degraded environment represents a potentially degraded tourism industry. If the local environment appears dirty and/or poorly managed, it threatens not only local livelihoods, but individuals' self-images as well. One respondent described the village's dependency this way: "And here in the village, the only source of work is tourism so, if tourists stop coming, we won't have work so . . . we have to take care of the nature because if we don't, they will stop coming. This is what I've seen since I've been here. Tortuguero is dependent on tourism and nature, the things that are here" (R55 2004). Due to the intimate ties between residents and the ecotourism product, identity is also at stake. For example, the village and its inhabitants are viewed as attractions for tourists.

Third, building on recent works by others (eg Khan 2002; Peña 2002, 2005; Zebich-Knos 2008), we demonstrate the utility of using EJ to understand environmental impacts of ecotourism in general, and the solid waste crisis in Tortuguero specifically. At first, it seems odd to contemplate environmental injustices in ecotourism settings, given ecotourism's explicit agenda of both conserving natural resources and contributing to local development. However, our case study is one in which residents identify the solid waste crisis as an environmental injustice, *despite* the benefits of ecotourism, and *despite* the village's relative wealth in its region. While the benefits of ecotourism are real, numerous, and much appreciated in Tortuguero, this does not change the perceived injustice of the solid waste crisis. As such, this case study is a good example of how "the compensation mentality", where economic benefits are assumed to overwhelm any negative impacts of a particular industry, remains flawed, even in newer, "greener" industries like ecotourism.

It seems especially counter-intuitive to raise the issue of EJ in Tortuguero, a renowned turtle conservation and ecotourism "success" in which the community is portrayed as playing a central role (Safina 2006; Smith 2005; Troëng and Rankin 2005). Our work focuses on different aspects of the story of ecotourism in Tortuguero; it draws attention to the tensions and conflicts that can and do exist between wildlife conservation needs, the aesthetic needs of ecotourism, and the human development needs, despite whatever successes can be measured. The EJ framing of our case study also exposes some of the problematic philosophical underpinnings of ecotourism in practice. Whereas conventional EJ industries may prioritize profit or extra-local development needs over the local environmental, health, and social wellbeing of marginalized communities, ecotourism has the potential to take that one undesirable step further. Conservation-driven ecotourism has the potential to place the needs of certain conservation goals (eg preserving biodiversity), or certain species (eg endangered sea turtles) over and above the needs of local communities, further contributing to their marginalization. The environmental injustices associated with the solid wastes crisis suggest that the "successful" reputation of ecotourism in Tortuguero is somewhat bio-, and more specifically, species-centric: if both turtle nesting numbers and the turtle tourism revenues are increasing, ecotourism is seen as "working" (Troëng 2004; Troëng and Drews 2004; Troëng and Rankin 2005). Focusing on turtles and turtle conservation as indicators of success is problematic because it masks issues arising in other domains of the environment, such as waste management. In Tortuguero, the focus of policy level and management-related efforts on turtle nesting and turtle viewing is sometimes criticized by residents because it is seen as coming at the expense of other important environmental issues that respondents view as equally if not more important than turtle conservation (eg respondents discuss illegal deforestation; poaching of other animals; water and soil pollution in Meletis 2007).

Finally, solid waste management, especially on the beach, should be of greater concern in turtle conservation efforts due to the direct connections between solid waste and turtle fitness, if nothing else. Waste buried on the beach is often exposed and re-exposed at the surface through wave action and bank erosion, where it decomposes and has the potential to leach hazardous materials or attract bacteria that could endanger turtle egg clutch health. Solid waste that ends up in the water also represents potential threats such as entanglement and ingestion-related problems (Derraik 2002; Özdilek et al 2006). Furthermore, turtle-watching ecotourists expect to share in pristine environments with the charismatic "main attractions"—sea turtles in this case. With respect to the ecotourism-associated solid waste crisis in Tortuguero, residents, tourists, and the turtles themselves all have the potential to "lose out" as a result of the industry's and the community's inability to manage solid wastes in the village. The reluctance of the local government, the Park, and the lodges to contribute to local waste management on a steady basis is seen as unfair and unjust by respondents because the "community of Tortuguero" remains marginalized, without the resources and support required to manage the negative environmental impacts that its members must live with but did not necessarily consent to.

Acknowledgements

This research was supported by the Social Science & Humanities Research Council of Canada; Duke University; the Mary Derrickson Dissertation Year Fellowship (Duke Marine Lab); the Oak Foundation and the John Hope Franklin Dissertation Working Group (Duke); the Tinker Melon Foundation & the Latin American & Caribbean Studies Center (Duke); the CAG; the AAG; the International Sea Turtle Society; and the University of Western Ontario. We acknowledge "Campbell lab members", Matthew Godfrey, the editors, and three reviewers for providing valuable comments. We thank the Caribbean Conservation Corporation for in-kind support and friendship, and most of all, we thank the people of Tortuguero.

Endnotes

[1] Unless they specifically asked to be identified, respondents are identified by alphanumeric codes in this chapter. We used additional wording in square brackets in some respondent quotes in order to clarify implied meanings, or to insert explanatory words not evident in de-contextualized quotes; the content remains that of the respondents.

[2] In this chapter, we use the term resident loosely; we define all people living in Tortuguero at the time as residents, regardless of their legal status.

[3] These authors use political ecology and political economy but focus on concepts central to EJ: race, class, and access to environmental goods.

[4] Serious infrastructure and servicing issues remain, however. For example, Tortuguero's schools are notoriously poor in quality and have staffing difficulties—the isolated location complicates staffing and staff retention.

[5] Despite some methodological shortcomings (eg waste was only measured over 4 days; waste was only quantified by weight, not by volume; the waste received from

village-based tourism businesses was not segregated but was instead counted as community-generated solid waste; only three of the lodges were bringing waste to the plant at the time), the waste audit offers interesting suggestions regarding waste generation in and around the village (eg lodges).

[6] Since 2004, waste management has improved in Tortuguero. A resident entrepreneur has taken over, and the plant is considered "reborn", with new equipment and its own boat. The new system also makes payment for waste collection and disposal services mandatory by tacking waste bills onto household water bills, raising ethical and legal questions (Sherwood 2007). The long-term success of this new leadership has yet to be determined, however, and many challenges lie ahead. Many residents would like to see the plant moved out of the center of town, and there are continued calls for increased government support and some kind of direct revenue contributions from tourism.

[7] There are exceptions, however. For example, some tour guides go boating (eg short canoe trips) in the canals for their own enjoyment, engaging in a direct use of the park.

[8] Until 2002, the village relied mainly on five shared "public phones" housed in village buildings. Within the last 7 years, dozens of public phone booths and hundreds of residential lines have been installed, in addition to a cell tower that relays signals to the hundreds of cell phones used locally.

[9] There are exceptions, however, such as the siting of "dirty industries" in The Tar Ponds (Sydney, Canada). Most employed residents were either directly or indirectly employed by these "dirty" industries, and financially benefited from their presence (eg Haalboom et al 2006).

[10] There are exceptions; some studies do identify heterogeneous and sometimes opposing views on such projects within communities (eg Bradshaw 2003).

References

Adeola F O (2000) Cross-national environmental injustice and human rights issues. *The American Behavioral Scientist* 43(4):686–706

Akama J S (1996) Western environmental values and nature-based tourism in Kenya. *Tourism Management* 17(8):567–574

Alexander J and McGregor J (2000) Wildlife and politics: CAMPFIRE in Zimbabwe. *Development and Change* 31(3):605–627

Alston D and Brown N (1993) Global threats to people of color. In R Bullard (ed) *Confronting Environmental Racism. Voices from the Grassroots* (pp 179–194). Boston: South End Press

Ateljevic I and Doorne S (2005) Dialectics of authentication: Performing "exotic otherness" in a backpacker enclave of Dali, China. *Journal of Tourism and Cultural Change* 3(1):1–17

Barrera L M (2003) "Institutions and conservation: Exploring the effects of social difference." Unpublished Master's thesis, University of Cambridge

Belkhir J A and Adeola F O (1997) Environmentalism, race, gender & class in global perspective. *Race, Gender & Class* 5(1):4–12

Benford R (2005) The half-life of the environmental justice frame: Innovation, diffusion, and stagnation. In D N Pellow and R J Brulle (eds) *Power, Justice and the Environment.* Cambridge, MA: MIT Press

Bhattarai K, Conway D and Shrestha N (2005) Tourism, terrorism and turmoil in Nepal. *Annals of Tourism Research* 32(3):669–688

Boo E (1990) *Ecotourism: The Potentials and Pitfalls.* Vols 1 and 2. Baltimore, MD: World Wildlife Fund

Bradshaw B (2003) Questioning the credibility and capacity of community-based resource management. *The Canadian Geographer* 47(2):137–150

Britton S G (1982) The political economy of tourism in the third world. *Annals of Tourism Research* 9(3):331–358

Brockington D (2004) Community conservation, inequity and injustice: Myths of power in protected area management. *Conservation & Society* 2(2):411–432

Brown K, Turner R K, Hameed H and Bateman I (1997) Environmental carrying capacity and tourism development in the Maldives and Nepal. *Environmental Conservation* 24(4):316–325

Bryant R L and Goodman M K (2004) Consuming narratives: The political ecology of "alternative" consumption. *Transactions of the Institute of British Geographers NS* 29:344–366

Bullard R D (1990) *Dumping in Dixie: Race, Class and Environmental Quality.* Boulder, CO: Westview Press

Bullard R D (1993) *Confronting Environmental Racism: Voices from the Grassroots.* Boston, MA: South End Press

Bullard R D (1994) Overcoming racism in environmental decision-making. *Environment* 36(4):10–20, 39–44

Bullard R (2001) Environmental justice in the 21st century: Race still matters. *Phylon* 49(3/4):151–171

Bullard R D and Johnson G S (2000) Environmental justice: Grassroots activism and its impact on public policy decision making. *Journal of Social Issues* 56(3):555–578

Camacho M A (2003) Informe sobre la gestion de manejo desechos solidos en la communidad de Tortuguero. San Jose: Costa Rica

Campbell L M (1999) Ecotourism in rural developing communities. *Annals of Tourism Research* 26(3):534–553

Campbell L M (2002a) Conservation narratives and the received wisdom of ecotourism: Case studies from Costa Rica. *International Journal of Sustainable Development* 5(3):300–325

Campbell L M (2002b) Conservation narratives in Costa Rica: Conflict and co-existence. *Development and Change* 33(1):29–56

Campbell L M (2003) Contemporary culture, use, and conservation of sea turtles. In P L Lutz, J A Musick and J Wyneken (eds) *The Biology of Sea Turtles* (pp 307–338). Boca Raton, FL: CRC Press

Campbell L M (2007) Local conservation practice and global discourse: A political ecology of sea turtle conservation. *Annals of the Association of American Geographers* 97(2):313–334

Campbell L M, Gray N J and Meletis Z A (2008) Political ecology perspectives on ecotourism to parks and protected areas. In K S Hanna, D A Clark and D S Slocombe (eds) *Transforming Parks and Protected Areas: Policy and Governance in a Changing World* (pp 200–221). New York and London: Routledge Taylor and Francis Group

Čapek S M (1993) The "environmental justice" frame: A conceptual discussion and an application. *Social Problems* 40(1):5–24

Caribbean Conservation Corporation (2003) Tortuguero National Park, Costa Rica, "Region of the turtles". Gainsville, FL. http://www/cccturtle.org/tortNP.htm Accessed 11 August 2006

Caribbean Conservation Corporation (2007) Tortuguero National Park, Costa Rica, "Region of the turtles". Gainsville, FL. http://www.cccturtle.org/volunteer-research-programs.php?page=tortNP Accessed 24 June 2006

Carrier J G and Macleod D V L (2005) Bursting the bubble: The socio-cultural context of ecotourism. *Journal of the Royal Anthropological Institute* 11:315–334

Carruthers D V (ed) (2008) *Environmental Justice in Latin America*. Cambridge and London: The MIT Press

Cater E (2002) Spread and backwash effects in ecotourism: Implications for sustainable development. *International Journal of Sustainable Development* 5(3):265–281

Cater E (2006) Ecotourism as a Western construct. *Journal of Ecotourism* 5(1&2):23–39

Ceballos-Lascuráin H (1996) Tourism, ecotourism and protected areas: The state of nature-based tourism around the world and guidelines for its development. Gland, Switzerland: IUCN

Charnley S (2005) From nature tourism to ecotourism? The case of the Ngorongoro Conservation Area, Tanzania. *Human Organization* 64(1):75–88

Chavis B J (1993) Foreword. In R D Bullard (ed) *Confronting Environmental Racism. Voices from the Grassroots* (pp 3–5). Boston, MA: South End Press

Cock J and Fig D (2002) From colonial to community-based conservation. In D A McDonald (ed) *Environmental Justice in South Africa* (pp 131–155). Athens, OH: Ohio University Press and Cape Town: University of Cape Town Press

Cohen E (1978) The impact of tourism on the physical environment. *Annals of Tourism Research* April/June:215–237

Colten C E and Dilsaver L M (2005) The hidden landscape of Yosemite National Park. *Journal of Cultural Geography* 22(2):27–50

Comfort S (2002) Struggle in Ogoniland. Ken Saro-Wiwa and the cultural politics of environmental justice. In J E Adamson, M M Evans and R Stein (eds) *The Environmental Justice Reader* (pp 229–264). Tucson, AZ: University of Arizona Press

De Haro A and Tröeng S (2006) *Report on the 2005 Green Turtle Program at Tortuguero, Costa Rica*. San Pedro, Costa Rica: Caribbean Conservation Corporation

De Oliviera J A P (2005) Tourism as a force for establishing protected areas: The case of Bahia, Brazil. *Journal of Sustainable Tourism* 13(1):24–49

Derraik J G B (2002) The pollution of the marine environment by plastic debris: A review. *Marine Pollution Bulletin* 44:842–852

Duffy R (2002) *A Trip Too Far*. London: Earthscan

Edwards N (2002) Radiation, tobacco, and illness in Point Hope, Alaska: Approaches to the "facts" in contaminated communities. In J Adamson, M M Evans and R Stein (eds) *The Environmental Justice Reader* (pp 105–124). Tucson, AZ: University of Arizona Press

EQUATIONS (2008) Call for Action—No more holidays from accountability! EQUATIONS statement on World Tourism Day, Bangalore. Email sent to Tourism Geography listserv (IGUST)

Evans S (1999) *The Green Republic. A Conservation History of Costa Rica*. Austin, TX: University of Texas Press

Fauzi A and Buchary E A (2002) A socioeconomic perspective of environmental degradation at Kepulauan Seribu marine national park, Indonesia. *Coastal Management* 30:167–181

Fletcher T H (2003) *From Love Canal to Environmental Justice. The Politics of Hazardous Waste on the Canada–US Border*. Peterborough: Broadview Press

Floyd M F and Johnson C Y (2002) Coming to terms with environmental justice in outdoor recreation: A conceptual discussion with research implications. *Leisure Sciences* 24(1):59–77

Garcia R (2002) Equal access to California's beaches. *Second National People of Color Environmental Leadership Summit—Summit II*. Resource Paper Series. Washington DC: Summit II National Office

Geisler C (2003) Your Park, my Poverty. using impact assessment to counter the displacement effects of environmnetal greenlining. In J M Belsky, P R Wilshusen, C L Fortwangler and P C West (eds) *Contested Nature: Promoting International*

Biodiversity with Social Justice in the Twenty-first Century (pp 217–229). New York: State University of New York Press

Goldman B A (1996) What is the future of environmental justice? *Antipode* 28(2):122–141

Goodier R (2005) Green sea turtles massacred on Caribbean coast. *The Tico Times* 23 August. http://www.ticotimes.net/dailyarchive/2005_08/daily_08_23_05.htm Accessed 15 April 2009

Goodman M K (2004) Reading fair trade: Political ecology imaginary and the moral economy of fair trade foods. *Political Geography* 23(7):891–915

Gray N (2002) Unpacking the baggage of ecotourism: Nature, science and local participation. *The Great Lakes Geographer* 9(2):11–21

Gregory M R (1999) Plastics and South Pacific Island shores: Environmental implications. *Ocean & Coastal Management* 42(6):603–615

Grossberg R, Treves A and Naughton-Treves L (2003) The incidental ecotourist: Measuring visitor impacts on endangered howler monkeys at a Belizean archaeological site. *Environmental Conservation* 30(1):40–51

Haalboom B, Elliott S J, Eyles, J and Muggah H (2006) The risk society at work in the Sydney "Tar Ponds". *The Canadian Geographer* 50(2):227–241

Halder B (2003) "Ecocide and genocide" explorations of environmental justice in Lakota Sioux country. In D G Anderson and E Berglund E (eds) *Ethnographies of Conservation. Environmentalism and the Distribution of Privilege* (pp 101–118). New York: Berghahn Books

Hall C M (1994) *Tourism and Politics. Policy, Power, and Place*. Chichester: John Wiley & Sons

Hall M C (2003) Politics and place: An analysis of power in tourism communities. In S T Singh, J Dallen and R K Dowling (eds) *Tourism in Destination Communities* (pp 99–113). Wallingford, UK: CABI Publishing

Hamilton C (1993) Confronting environmental racism. Voices from the grassroots. In R D Bullard and B J Chavis (eds) *Confronting Environmental Racism* (pp 63–75). Boston, MA: South End Press

Harrison E (2009) Email confirmation of TNP tickets sold in 2008: 134, 255 (number from Ministerio Del Ambiente Y Energia (MINAE))

Harrison E, Troëng S and various research assistants (2004) Report on the 2004 Green Turtle Program at Tortuguero, Costa Rica. San Pedro, Costa Rica: Caribbean Conservation Corporation

Harvey D (1996) *Justice, Nature & The Geography of Difference*. Oxford, UK: Blackwell Publishing

Heyman J M (2005) The political ecology of consumption. In S Paulson and L L Gezon (eds) *Political Ecology across Spaces, Scales, and Social Groups*. New Brunswick, NJ: Rutgers University Press

Heynen N, Perkins H A and Roy P (2008) The political ecology of uneven green space *Urban Affairs Review* 42(1):3–25

Hillery M, Nancarrow B, Griffin G and Syme G (2001) Tourist perceptions of environmental impact. *Annals of Tourism Research* 28(4):853–867

Homewood K M (1991) *Maasailand Ecology: Pastoralist Development and Wildlife Conservation in Ngorongoro, Tanzania*. Cambridge, UK: Cambridge University Press

Honey M (1999) *Ecotourism and Sustainable Development. Who Owns Paradise?* Washington DC: Island Press

Hughes D M (2005) Third nature: Making space and time in the Great Limpopo Conservation Area. *Cultural Anthropology* 20(3):157–184

Hughes M and Morrison-Saunders A (2003) Visitor attitudes toward a modified natural attraction. *Society and Natural Resources* 16(3):191–203

Jarquin O and Gayle A (2004) Análysis de Situación de Salud Para el Sector. Tortuguero: EBAIS (Regional medical team assigned to seven villages in Tortuguero's municipality)

Johnston A M (2003) Self-determination: Exercising indigenous rights in tourism. In S T Singh, J Dallen, R K Dowling (eds) *Tourism in Destination Communities* (pp 115–133). Wallingford, UK: CABI Publishing

Khan F (2002) The roots of environmental racism and the rise of environmental justice in the 1990s. In D A Mcdonald (ed) *Environmental Justice in South Africa* (pp 15–49). Athens, OH: Ohio University Press and Cape Town: University of Cape Town Press

Kirby A (2004) Islands face major waste problem. Jeju, Korea, BBC News. http://news.bbc.co.uk/go/pr/fr/-12/hi/science/nature/3580253.stm Accessed 1 April 2004

Kuletz V (2002) The movement for environmental justice in the Pacific Islands. In J E Adamson, M M Evans and R Stein (eds) *The Environmental Justice Reader* (pp 125–142). Tucson, AZ: University of Arizona Press

Kuo N W and Yu Y H (2001) An investigation of the environmental loads of Shei-Pa National Park in Taiwan. *Environmental Geology* 40(3):312–316

Lee C (1993) Beyond toxic wastes and race. In R Bullard (ed) *Confronting Environmental Racism. Voices from the Grassroots* (pp 41–52). Boston, MA: South End Press

Lee D N B and Snepenger D J (1992) An ecotourism assessment of Tortuguero, Costa Rica. *Annals of Tourism Research* 19(2):367–370

Lefever H G (1992) *Turtle Bogue. Afro-Caribbean Life and Culture in a Costa Rican Village.* Selingsgrove: Susquehanna University Press and Cranbury, NJ: Associated University Press

Levi J M (1999) Hidden transcripts among the Rarámuri: Culture, resistance, and interethnic relations in northern Mexico. *American Ethnologist* 26(1):90–113

Lindberg K, Enriquez J and Sproule K (1996) Ecotourism questioned. Case studies from Belize. *Annals of Tourism Research* 23(3):543–562

Lyon S (2006) Evaluating fair trade consumption, politics, defetishization and producer participation. *International Journal of Consumer Studies* 30(5):452–464

Manokha I (2004) Modern slavery and Fair Trade products: Buy one and set someone free. In C Van Den Anker (ed) *The Political Economy of New Slavery* (pp 217–234). New York: Palgrave Macmillan

Martinez-Alier J (2003) Mining conflicts, environmental justice and valuation. In J Agyeman, R D Bullard and B Evans (eds) *Just Sustainabilities—Development in an Unequal World* (pp 201–228). Cambridge, MA: MIT Press

Maschesky W (2003) Environmental justice in Scotland—just words? A view from the outside (for) Friends of The Earth Scotland. http://www.foe-scotland.org.uk/nation/ej_a_view_from_outside.pdf Accessed 1 September 2003)

McDonald D A E (2002) *Environmental Justice in South Africa.* Athens, OH: Ohio University Press and Cape Town: University of Cape Town Press

McDonald M and Wearing S (2003) Reconciling communities' expectations of Ecotourism: Initiating a planning and education strategy for the Avoca Beach Rock Platform. In B Garrod and J C Wilson (eds) *Marine Ecotourism. Issues and Experiences* (pp 155–170). Clevedon: Channel View Publications

McMinn S and Cater E (1998) Tourist typology. Observations from Belize. *Annals of Tourism Research* 23(3):675–699

Meletis Z A (2004) *La planta esta funcionando?* [The plant is working?]. Costa Rica and the USA, 22 minutes

Meletis Z A (2007) "Wasted visits? Ecotourism in theory vs. practice, at Tortuguero, Costa Rica." Unpublished PhD thesis, Duke University

Meletis Z A and Campbell L M (2007) Call it consumption! (Re)conceptualizing ecotourism as consumption and consumptive. *Geography Compass* 1(4):850–870

Ministerio Del Ambiente Y Energia (MINAE) (no date) Area de Conservación Tortuguero (ACTo). San Jose: Costa Rica

Moberg M (2005) Fair Trade and eastern Caribbean banana farmers: Rhetoric and reality in the anti-globalization movement. *Human Organization* 64(1):4–15

Moore S A (2008) Waste practices and politics: The case of Oaxaca, Mexico. In D V Carruthers (ed) *Environmental Justice in Latin America* (pp 119–136). Cambridge, MA: MIT Press

Mowforth M and Munt I (1998) *Tourism and Sustainability. New Tourism in the Third World*. London: Routledge

Nelson V (2005) Representation and images of people, place and nature in Grenada's tourism. *Geografiska Annaler* 87 B(2):131–143

Njeru J (2006) The urban political ecology of plastic bag waste problem in Nairobi, Kenya. *Geoforum* 37(6):1046–1058

Norton A (1996) Experiencing nature: The reproduction of environmental discourse through safari tourism in East Africa. *Geoforum* 27(3):355–373

Nygren A (2003) Nature as contested terrain. Conflicts over wilderness protection and local livelihoods in Rio San Juan, Nicaragua. In D G Anderson and E Berglund (eds) *Ethnographies of Conservation. Environmentalism and the Distribution of Privilege* (pp 33–49). New York: Berghan Books

Orams M B (1999) *Marine Tourism. Development, Impacts and Management*. London: Routledge

Özdilek H G, Yalçin-Özilek S, Ozaner F S and Sönmez B S (2006) Impact of accumulated beach litter on *Chelonia mydas* L. 1758 (Green Turtle) hatchlings of the Samandağ coast, Hatay, Turkey. *Fresenius Environmental Bulletin* PSP Volume 15(2):1–9

Peisch S J (no date) Pesticides in the Barrio: The case of Guayanilla, Puerto Rico. *EJRC- Voices From the Grassroots*. http://www.ejrc.cau/POCEG_03.PDF Accessed 29 March 2007

Pellow D N (2000) Environmental inequality formation: Toward a theory of environmental injustice. *The American Behavioral Scientist* 43(4):581–601

Pellow D N and Brulle R J (2005) *Power, Justice, and the Environment*. Cambridge, MA: MIT Press

Peña D G (2002) Endangered landscapes and disappearing peoples? Identity, place, and community in ecological politics. In J E Adamson, M M Evans and R Stein (eds) *The Environmental Justice Reader* (pp 58–81). Tucson, AZ: University of Arizona Press

Peña D G (2005) Autonomy, equity, and environmental justice. In D N Pellow and R J Breulle (eds) *Power, Justice and the Environment* (pp 131–151). Cambridge, MA: MIT Press

Perkins H A and Heynen N (2004) Inequitable access to urban reforestation: The impact of urban political economy on housing tenure and urban forests. *Cities* 21(4):291–299

Peskin J (2002) "Local guides' attitudes toward ecotourism, sea turtle conservation, and guiding in Tortuguero, Costa Rica." Unpublished Master's thesis, University of Florida

Place S (1991) Nature tourism and rural development in Tortuguero. *Annals of Tourism Research* 18(1):186–201

Pleumarom A (1999) Eco-tourism: An ecological and economic trap for Third World Countries. Third World Network. http://twnside.org.sg Accessed 29 March 2007

Porter R and Tarrant M A (2001) A case study of environmental justice and federal tourism sites in southern Appalachia: A GIS application. *Journal of Travel Research* 40(1):27–40

Pulido L (1996) A critical review of the methodology of environmental racism research. *Antipode* 28(2):142–159

Rees W E and Westra L (2003) When consumption does violence: Can there be sustainability and environmental justice in a resource-limited world? In J Ageyman, R D Bullard and B Evans (eds). *Just Sustainabilities. Development in an Unequal World* (pp 99–124). Cambridge, MA: MIT Press

Ryan C, Hughes K and Chirgwin S (2000) The gaze, spectacle and ecotourism. *Annals of Tourism Research* 27(1):148–163

Safina C (2006) *Voyage of the Turtle. In Pursuit of the Earth's Last Dinosaur*. New York: Henry Holt and Company

Scheyvens R (1999) Ecotourism and the empowerment of local communities. *Tourism Management* 20:245–249

Scheyvens R (2002) *Tourism for Development. Empowering Communities*. Harlow, Essex: Prentice Hall, Pearson Education

Schlosberg D (1999) *Environmental Justice and the New Pluralism*. Oxford and New York: Oxford University Press

Sherwood D (2007) Recycling center cleans up Tortuguero area. *The Tico Times* San Jose, Costa Rica 13 July

Silman R (2002) Costa Rica provides a world example for sea turtle protection (Caribbean Conservation Press Release). Gainsville, FL: Caribbean Conservation Corporation. http://cccturtle.org/pressreleases.php?page=n_cr-passes-new-law Accessed 1 April 2009

Simpson A (2002) Who hears their cry? African American women and the fight for environmental justice in Memphis, Tennessee. In J Adamson, M M Evans and R Stein (eds) *The Environmental Justice Reader* (pp 82–104). Tucson, AZ: University of Arizona Press

Sindiga I (1995) Wildlife-based tourism in Kenya: Land use conflicts and government compensation policies over protected areas. *Journal of Tourism Studies* 6(2):45–55

Smith D (2005) *La Comunidad. Tortuguero, Costa Rica. Una pelicula sobre la gente y la naturaleza de Tortuguero*. Costa Rica and USA: Associacion CAVU

Smith M and Duffy R (2003) *The Ethics of Tourism Development*. London: Routledge

Smith N (2006) There's no such thing as a natural disaster. *Understanding Katrina*. http://understandingkatrina.ssec.org/Smith/ Accessed 31 May 2007

Stern C J, Lassoie J P, Lee D R and Deshler D J (2003) How "eco" is ecotourism? A comparative case study of ecotourism in Costa Rica. *Journal of Sustainable Tourism* 11(4):322–347

Talbot J M (2004) *Grounds for Agreement. The Political Economy of the Coffee Commodity Chain*. London: Rowman & Littlefield

Tarmann K F (2002) The Fair Trade Coffee Movement: Norm change or niche marketing? Woodrow Wilson Department of Politics. Arlington, VA: University of Virginia

Tarrant M A and Cordell H K (1999) Environmental justice and the spatial distribution of outdoor recreation sites: An application of geographic information systems. *Journal of Leisure Research* 31(1):18–34

The International Ecotourism Society (no date) Ecotourism—definitions and principles. https//:ecotourism.org Accessed 12 June 2006

Thrupp L A (1990) Environmental initiatives in Costa Rica: A political ecology perspective. *Society and Natural Resources* 3:243–256

Torres R and Skillicorn P (2004) Montezuma's revenge: How sanitation concerns may injure Mexico's tourist industry. *Cornell Hotel and Restaurant Administration Quarterly* 45(3):132–144

Trist C (1999) Recreating open space: Recreational consumption and representation of the Caribbean marine environment. *Professional Geographer* 51(3):376–387

Troëng S (2004) Five decades of sea turtle conservation, monitoring and research at Tortuguero. Presentation given at Tortuga Lodge, 8 August 2004. Tortuguero, Costa Rica

Troëng S and Drews C (2004) Money talks. Economic aspects of marine turtle use and conservation. Gland, Switzerland: WWF-International

Troëng S and Rankin E (2005) Long-term conservation efforts contribute to positive green turtle *Chelonias Mydas* nesting trend at Tortuguero, Costa Rica. *Biological Conservation* 121(1):111–116

Urry J (1995) *Consuming Places*. London: Routledge

Urry J (2002) *The Tourist Gaze*. London: Sage Publications

US General Accounting Office (1983) Siting of hazardous waste landfills and their correlation with racial and economic status of surrounding communities. Washington DC: Government Printing Office

Walker G and Bulkeley H (2006) Editorial: Geographies of environmental justice. *Geoforum* 37(5):655–659

Weaver D B (1998) *Ecotourism in the Less Developed World*. Wallingford, UK: CABI Publishing

Weaver D B (1999) Magnitude of ecotourism in Costa Rica and Kenya. *Annals of Tourism Research* 26(4):792–816

Weaver D B (2001) *Ecotourism*. Milton: John Wiley & Sons Australia

Weaver D B (2002) The evolving concept of ecotourism and its potential impacts. *International Journal of Sustainable Development. Special Issue: Ecotourism.* 5(3):251–264

West P C, Fortwangler C L, Agbo V, Simsik M and Sokpon N (2003) The political economy of ecotourism. Pendjari National Park and ecotourism concentration in northern Benin. In S R Brechin, P R Wilshusen, C L Fortwangler and P C West (eds) *Contested Nature: Promoting International Biodiversity with Social Justice in the Twenty-first Century* (pp 103–115). New York: State University of New York Press

Wight P A (1996) *North American Ecotourists: Market Profile and Trip Characteristics*. Thousand Oaks, CA: Sage Publications

Wu C and Wang S (2002) Environment, development and human rights in China: A case study of foreign waste dumping. In L Zarsky (ed) *Human Rights and The Environment*. London: Earthscan

Zebich-Knos M (2008) Ecotourism, park systems, and environmental justice in Latin America. In D V Carruthers (ed) *Environmental Justice in Latin America* (pp 185–211). Cambridge, MA: MIT Press

Chapter 7

Assembling Justice Spaces: The Scalar Politics of Environmental Justice in North-east England

Karen Bickerstaff and Julian Agyeman

Introduction

The US environmental justice (EJ) movement emerged in the late 1970s as a mode of activism which embodied a vision of a just society that explicitly linked to a history of civil rights struggles (Agyeman and Evans 2004, Capek 1993, Sandweiss 1998, Taylor 2000). It is characterised by a discourse (or *frame*) that identifies the deliberate targeting of particular communities (of colour and/or low income) as the main source of environmental injustice. Spurred by such activism, EJ scholars have, since the 1980s, sought to demonstrate empirically that environmental burdens, such as proximity to hazardous sites, are inequitably distributed both spatially and socially. This literature, which Williams (1999) terms the "first wave" of EJ scholarship,[1] is primarily concerned with revealing the extent of environmental inequities throughout America. This identity-politics framing of EJ is, as Faber (2005) argues, the dominant orientation of the US EJ Movement— and one that has succeeding in making issues of race, class, culture, and gender integral to the discourse and politics of environmentalism. In the USA, the clearest victories have been won at the level of specific community struggles, benefiting from the power of the wider movement to focus strength in a local context or a definable space. There has also been some success in effecting a series of federal and state policy actions (Bullard and Johnson 2000, Williams 1999)—though the force of these has been questioned by some authors (Holifield 2004, Pellow and Brulle 2005).

In the UK, the range and depth of research with a distributional focus is more limited than in the USA (Walker et al 2005). A growing body of academic and praxis research has pointed to national similarities in the social and spatial contours of environmental hazards (such as atmospheric chemical releases from large-scale industrial processes, landfills, and sites registered under Control of Major Accident Hazard regulations), with poorer communities, in particular, disproportionately impacted by local sources of environmental harm. However, there is no vocabulary seeking to mobilise minority (ethnic) and low-income groups to compare with the racial politics of the USA (Agyeman and Evans 2004; Dobson 1998). Grassroots politics have not emerged in as strong a form and the UK is much further from recognising EJ as an intrinsic part of policy and decision-making processes.

Such contrasts in the politics of EJ arguably reflect fundamental differences in patterns of land use and the political, legislative, economic, and administrative frameworks and cultures of each country (see Walker et al 2005). Where EJ has become a salient concept for UK NGOs this is reflected in a set of ideas that emphasise issues of sustainability, social inclusion and the delivery of procedural environmental equity (Agyeman 2002; Boardman, Bullock and McLaren 1999; Friends of the Earth 2001; London Sustainability Exchange 2004; Stephens, Bullock and Scott 2001). What has been termed the "just sustainability" frame (Agyeman 2005; Agyeman, Bullard and Evans 2003) knits together an encompassing set of concerns associated with the multiple sites, forms and processes of injustice (Agyeman 2002; Walker and Bulkeley 2006) and has most strongly been articulated by public policy agencies, primarily at a national scale (DEFRA 2008; Environment Agency 2004, Health Protection Agency 2005).

There are a number of issues that such representations of (national) EJ experience and politics present for research concerned with the scalar dynamics of social mobilisation. Firstly, the literature (in particular that associated with Williams' first wave) is characterised by reified representations of scale—which are unable to accommodate multiplicity, change and, by implication, the socially and politically constructed nature of scale (Kurtz 2003; cf Paasi 2004). Other scholars, again seeking to establish the extent of environmental injustice, have posited a variety of credible explanations for unequal distributions that refer to social processes operating at a range of spatial scales— including political exclusion and market economics (Williams' second wave of EJ studies). Yet the problem of environmental inequity and its documentation (or fixing in space and time) remains inextricably bound up with a high level of spatial ambiguity, "with no indisputable rationale for favouring one scale of analysis over another" (Kurtz 2003:888; cf Walker et al 2005; Pulido 1996). Research therefore has to grapple with EJ as a multiscalar set of relationships of contradictions and dependencies (Debbané and Keil 2004:210).

Secondly, work more firmly rooted in the responses of affected communities has advocated (and even demonstrated) new forms of collectivity, social association and community around siting issues and risk exposures (Agyeman and Evans 2004; Bullard and Johnson 2000; Taylor 2000). Yet such calls can assume that communities, or groups within them, will act as coherent and equal units and as such embed a limited (and even universalistic) construction of the spatial politics of places. It is, for instance, clear that the dynamics of political mobilisation in populations exposed to technological hazards are far from unified (see, for instance, Freudenberg 1997; Kroll-Smith and Couch 1990). This argument also assumes that collective advocacy is *per se* appropriate and just—and as such claims will not be parochial or somehow inequitable at other scales of analysis. As Heynen (2003) points out, little attention has been given in this literature to understanding how socio-natural injustices at particular scales do not necessarily translate into injustices at other scales.

Finally, much research in the EJ field has addressed the collective framing of discourses which produce and maintain meanings—such as the identification of and active response to a social grievance—and in consequence provide movement participants with a common point of reference and identity at national (eg Agyeman 2002; Benford 2005; Taylor 2000) and also sub-national scales (eg Kurtz 2003). For such authors, frames are not static, reified entities but are continuously being constituted, contested, reproduced, transformed or replaced during the course of social movement activity. They are deployed to legitimate movement goals and campaigns by mobilising potential adherents and constituents, and demobilising any antagonists (Benford and Snow 2000). However, for the most part, such work pays scant attention to the active role of spatial constructs in framing practices (notable exceptions are Heynen 2003; Kurtz 2003; Towers 2000) and, given the foregrounding of discursive processes, steers away from recognising the material things that are simultaneously implicated in configuring the spaces of activism.

In this chapter we therefore contribute to research which has sought to attend to the spatial production and consequences of EJ activism. In doing so, our account takes as its focus a recent campaign in the north-east of England linking a local environmental group with the (inter)national NGO Friends of the Earth (FOE), to consider how scale(s) are constituted by and constitute EJ, how and why particular scales are invoked, and the effects and implications for our understanding of EJ. The first section sets out our theoretical approach to the scalar politics of EJ, followed by a discussion of the study area and the case study EJ group, based in Teesside. In the body of the chapter we address the spatial politics of an environmental controversy that centred precisely on the

Teesside area, and explore the analytical purchase of an assemblage perspective which draws together people, texts, machines, animals, devices and discourses in relations that collectively constitute—and scale—EJ. To conclude, and building upon this approach, we suggest future research avenues that we believe present a promising agenda for critical engagement with the production, scaling and politics of environmental (in)justice.

The Scalar Politics of Environmental Justice

Work in human geography on the politics of scale (for an extended discussion, see Delaney and Leitner 1997; Howitt 1993; Paasi 1991; MacLeod 1999) has addressed the contested ways in which scale is produced by social actors—and has already been a powerful influence on a small body of research seeking to account for the scalar trajectories of EJ activism (Kurtz 2003; Heynen 2003; Towers 2000; Williams 1999). For these authors, the very concept of environmental injustice precipitates a politics of scale—since locally experienced sources of pollution are inevitably rooted in political–economic relations and processes distributed across far-reaching spatial networks (Kurtz 2003; Towers 2000). Kurtz (2003), for instance, demonstrates a concern with the ways in which different actors invoke particular scales strategically in negotiations over the meaning, extent and responses to injustices. In a similar vein, Towers (2000) makes a useful distinction between "scales of meaning", the scale at which a problem is experienced and framed in political discourse, and "scales of regulation", which define landscapes administered by decision-making bodies. How individuals and social movement actors assemble frames—and the strategies they choose to deploy—will thus, in part at least, depend upon the structural opportunities and constraints present within a particular spatial and political context (cf Nicholls and Beaumont 2004). Kurtz (2003) similarly deploys the concept of scale frames to capture the specific discursive practices that construct meaningful linkages between the scale at which a social problem is experienced and the scale(s) at which it could be politically addressed or resolved. Through her analysis of the scale frames mobilised around a controversial siting proposal in Convent, Louisiana, Kurtz highlights different ways in which activists strategically mobilise geographic scale(s).

In this chapter we draw on these ideas about the scalar politics of EJ but, somewhat in contrast to Kurtz's analytical focus on frames, we seek to describe the scaling of activism through the assemblage of diverse entities. Developed initially by Bruno Latour, Michel Callon, and John Law, ANT provides a way of studying social and spatial orderings as contingently achieved through the enlistment or enrolment of human and non-human (eg material and animal) actors. Drawing on this work, there are a number of core ideas or principles of an assemblage perspective that we wish to pursue in the remainder of this chapter. Bennett (2005) refers to an assemblage as an ad hoc grouping, made up of many types of actors or actants, through which power is distributed—if unequally. Agency is thus conceived of as a relational effect of interaction with others, be that a human, a text, or an artefact (an idea evoked by the term actant). Crucially, if actors are to extend their influence or networks beyond what is physically immediate, the means have to be found to "act at a distance" by connecting to faraway events, places and people (Latour 1987; Murdoch 1997). This is primarily achieved through intermediaries— "anything passing between actors which defines the relationship between them" (Callon 1991:134). In this way material things begin to emerge as much more than simply the outcrops of human intention and action, for they actively frame, define and configure processes of interaction (Murdoch 1997:329). Finally, the relations between actors in an assemblage vary in stability, time–space extension and time–space form (Castree 2002:118; Rutland and Aylett 2008). Continuous labour and resources are therefore needed to enrol and control other actors as allies—to effectively produce and maintain an order.

Much of the current literature on EJ mobilisation, including recent work on the scalar politics of EJ, has attended to social ties and discursive resources—in effect to human agency. Yet, as Latour (2005:67) argues, attention (solely) to "discourse, structures or culture does not account for the way [power] exerts its grip over us and accounts for the unequal landscape". Our modest aim here, then, is to provide an exploration of how we might begin to engage with what Castree (2002) describes as shared capacities: a recognition of the non-human things that enable the social to endure over space and time—that build something "structural", whether social order, power or scale. This is not a call for the symmetrical treatment of agency (cf Castree 2002; Laurier and Philo 1999; Murdoch 1997) but rather for a (critical) engagement, among EJ researchers, with the implications of thinking and describing in terms of associations where both human and non-human actors have identities.

The bulk of research on the scalar politics of EJ has been situated within an explicitly US context (for an exception, see Debbané and Keil 2004 on Canada and South Africa). While there has been some discussion of collective framings of, and responses to, EJ in the UK (Agyeman 2002; Agyeman and Evans 2004), this limited body of work has not substantially engaged with theoretical ideas about the social, material and, crucially, spatial dynamics of politics and practice. Our purpose in this chapter is therefore to explore the (scalar) assemblage of EJ, through the lens of a major EJ campaign in the north-east of England.

Assembling Environmental Justice in North-east England: Justice for Teesside

The north-east is one of nine regions in England. As an old industrial region the socio-economic position of the north-east has been subject to a long-run decline due to the structural changes generated by de-industrialisation, faltering waves of branch plant-led reindustrialisation and somewhat limited tertiarisation (O'Brien, Pike and Tomancy 2004:64; cf Hudson 1998; Robinson 2002). The region has consequently fallen into a relatively marginal position in the national political economy (O'Brien, Pike and Tomancy 2004) and remains blighted by very high levels of multiple deprivation[2] (Health Protection Agency 2005; Hudson 2006; Walker et al 2003).

In this chapter, we are concerned with the formation and activities of a local campaign group based in the Teesside area of the north-east (see Figure 1), established and supported by the (inter)national environmental NGO FOE England, Wales and Northern Ireland. The initiative, referred to as the "Teesside Project" evolved out of FOE's Factory Watch Campaign in 1998, which published industrial pollution data online, enabling people to monitor factories in their own area

Figure 1: North-east England (source: North East Regional Information Partnership)

(Gilligan and Zagrovic 2004). Subsequent work compared the location of sites coming within the Integrated Pollution Control (IPC) system in England and Wales and local emissions data (principally chemicals recognised to be carcinogenic), with the Government's Index of Multiple Deprivation. The results provided initial evidence of a far from equitable distribution of harmful sources of pollution (FOE 2001). Teesside, a now-redundant administrative area that falls within the Tees Valley city-region,[3] was singled out as representing the extreme of environmental injustice. The analysis revealed that the Seal Sands area of Teesside housed the most industrial sites (17 in all). The average household income was just £6200—45% of the regional average income, or 36% of the national average (McLaren et al 1999).

Teesside was (and remains) a product of nineteenth century industrial expansion (Beynon, Hudson and Sadler 1994:11), and its growth was led by iron, steel, engineering and shipbuilding industries (1994:24). From the turn of the twentieth century, chemical industries (led by ICI) began their rise and, by the middle of the century, the Teesside area was among the most heavily industrialised regions of Western Europe. By the 1970s, however, the area's economy had entered a period of spiralling decline, caused partly by growing international competition. The Tees Valley still houses the largest integrated chemicals complex in the UK (Department for Communities and Local Government and Tees Valley Strategy Unit 2006) and roughly 12,000 homes lie within short distances (1 km) of Redcar Steel Complex and the petrochemical cluster at Wilton, Billingham and Seal Sands (Bush, Moffatt and Dunn 2001). The economy of "Teesside" is therefore one characterised by the double insecurity of economic decline and environmental pollution (Bush, Moffatt and Dunn 2001; Phillimore and Moffatt 1999).

Following recognition that deprived areas with high levels of industrial pollution were also the places with the lowest presence of FOE local groups, the "Teesside Project" was established to build local campaigning activity and capacity on EJ—initially resourced and supported by FOE. With a new programme of work on EJ it was also necessary for the organisation to demonstrate experience of working with affected communities (Gilligan and Zagrovic 2004). A commentary provided by members of the FOE team (Gilligan and Zagrovic 2004) sets out the following aims for the project:

- To increase local understanding of pollution threats and ways of influencing them
- To gain practical experience of empowering, and working hands-on with community groups.

- To have an impact on key decision-makers such as the Environment
  Agency, local authorities and MPs (eg to increase awareness of
  local concerns and achieve substantial reductions in levels of air
  pollution in communities living within the project area).

To deliver these aims, a community development worker was recruited
in 2001 and based in the town of Grangetown, situated adjacent to
the Wilton Petrochemical complex. Grangetown is ranked within the
top 10 most deprived wards (out of 8414) nationally (Tees Valley Joint
Strategy Unit 2000) and suffers the twin burdens of mass unemployment
and chronic ill health (BBC News 2004). Early work involved engaging
local communities in developing a common agenda, establishing local
networks and providing a "grassroots" office in Grangetown for advice,
information and a space to meet (Gilligan and Zagrovic 2004:8). The
Impact group was established in 2002—supported by, but independent
of, FOE. Early meetings, held in a local pub, attracted an "eclectic
mix of Green Party members, concerned residents and self-proclaimed
anarchists" (Every Action Counts 2008) from across the Tees Valley. By
2004 membership numbered 17. The group was explicitly concerned
with EJ in terms of "support[ing] the disadvantaged communities of
Teesside in reducing the impact of local pollution on environment
and quality of life" (Impact 2003). Given the aims set by FOE, initial
project activities focused on identifying local peoples' concerns about
industrial pollution, equipping Impact members with campaigning skills
and promoting engagement with the statutory environmental regulator
(the Environment Agency). In this way, the establishment and activities
of the group embedded an explicitly local framing of EJ in terms of both
the problem definition and the appropriate level of decision-making.

The FOE community worker played a crucial mediating role, not
only in enabling the nascent (and, as such, fragile) network of local
residents, but also in negotiating the sometimes uneasy relationship
with FOE (Scandrett 2005). The use of information and communication
technologies cannot be underestimated here—Impact put together a
website, leaflets and a popular internet discussion group (Every Action
Counts 2008). Indeed, Impact was described by a leading member of
the FOE team as having more of a virtual than a "real" presence: "[the
discussion] group has quite a few members and they used to post quite
a bit of information and discussion".[4] While formal meetings were
poorly attended, the internet and telephone kept campaigners across
Teesside—and further afield—coordinated, and facilitated the sharing
of information and research. Action thus emerged from associations,
distributed along the chain of humans and non-humans that held the
Impact network together.

Scaling Environmental Justice: Expansion and Contraction

The scalar ambitions of the Impact/FOE coalition shifted dramatically in 2003 when the Teesside area became the focus of a major public controversy, centring on what became known as the "Hartlepool ghost ships", which positioned the newly formed activist group and FOE centre stage. The Hartlepool ghost ships are part of the now aged ghost fleet of more than 100 ex-US Navy auxiliary vessels, administered by the US Department of Transport's Maritime Administration (MARAD) and moored on the James River in Virginia. In June 2003, Able UK, which owns a breaking yard for marine structures at Graythorp in Hartlepool, was awarded a £10.6 million contract to recycle the steel, and dispose of the pollutants, from 13 of these redundant ships. FOE, alerted by Impact to the possible local environmental threat, and buoyed by initial work in Teesside around industrial pollution, began a campaign to stop the ships being sent to Teesside.

With national campaigning expertise from FOE, and community pressure from Impact,[5] the "ghost ships" frame enrolled the ships as fragile assemblages of "dangerous toxic materials" (Juniper 2003, cited in FOE 2003d): "floating timebombs" (Guardian 2004; Williams 2003) whose physical integrity was "dangerously deteriorating" (Juniper 2003, cited in FOE 2003d). In so doing, the framing challenged the idea of the vessels as durable networks—suggesting that these ships were liable to catastrophic dis-assembly through corrosion and leakage of toxic materials: "asbestos, oil and PCBs" (FOE 2003a). The risk posed by the ships in situ would be greatly exacerbated should they be moved through "turbulent waters and buffeted by unpredictable weather systems across the thousands of nautical miles of rough seas of the North Atlantic" [Basel Action Network (BAN) 2003]. The implied threat was the sinking or breaching of the vessels in coastal waterways or on the high seas. The mobility and durability of PCBs in the marine environment, with the potential to alter the DNA and genetic integrity of species, presented an irreversible danger to animal and human bodies alike. The following statement from BAN (the Seattle-based campaign organisation aimed at confronting the global environmental injustice of toxic trade) conveys the threat of the ghost ships very squarely in terms of the agency of these chemical substances across time and space: "As PCBs are very stable they do not readily break down in the environment, and are able to persist for very long periods of time. PCBs can travel long distances in the air . . . PCBs ignore geographic barriers" (BAN 2003:25).

Additionally, an (in)justice frame was invoked, pointing to the inequitable risk exposure of Teesside residents—already heavily

burdened by a legacy of industrial pollution. In evidence to the House
of Commons Environment, Food and Rural Affairs Committee (House
of Commons hereafter), the local, Hartlepool, FOE group[6] made the
following points:

> There is a theory that: Where poverty is most concentrated, so are the
> poisons . . . Such levels of industrial air pollution have been linked to
> Teesside's relatively poor state of public health . . . we do not feel that
> such large scale, waste-generating and potentially hazardous ventures
> should be located in areas already blighted by the negative effects of
> industrial pollution . . . (House of Commons 2004:114)

The "ghost ships" campaign therefore appealed directly to a local
(Teesside) identity, centring on the legacy of pollution injustices and
stigma wrought on the area (by heavy industry and now by the USA),
with Hartlepool projected as an international dumping ground: "We've
got better things to do than break up shitty ships . . . We'll be looked on as
George Bush's toilet" (Mike Turner, Labour councillor in Seaton ward,
cited in Vidal 2003). Indeed, in expanding the network and enrolling
other social allies, FOE campaigners directly linked this local sense of
injustice to wider scales of regulation and, in particular, a perceived
failure of international enforcement of the proximity principle. It was
argued that waste disposal and recycling should be dealt with as close
to source as possible—ie that the ships should never have left the USA,
which had a moral obligation (as well as the skills and capacity) to deal
with its own waste (FOE 2004).

The organisational resources that FOE was able to wield were
considerable and, at the request of Impact members, a legal objection,
through an application to the High Court for a Judicial Review, was
launched in October 2003 (FOE 2003c) with the aim of preventing
the ships docking in Hartlepool. FOE claimed that the Environment
Agency had acted illegally in granting a modified waste management
licence on the basis that a proper environmental assessment had not
been undertaken (FOE 2003c). Before the case against the Environment
Agency had been heard, the regulator accepted that the original licence
could no longer stand and as such it would not contest FOE's application
to quash the modification, which was subsequently ruled unlawful and
annulled.

In a linked (and simultaneous) action, aimed at preventing the ships
leaving the USA, the BAN, the Sierra Club, and Earthjustice brought
a lawsuit against the authorities in the USA for failures in the way
this 'export of pollution' had been handled. Declarations supporting
the environmental injustice claims of the BAN challenge were made
by members of Impact. The judge granted a partial restraining order—
allowing four ships to proceed to the UK, while blocking the remaining

nine (BAN 2003). At the national (and international) scale of regulation, FOE were able—via a chain of (absent) licences and permissions, formal inquiries and written submissions—to translate their goals, forcibly effecting the (temporary) enrolment of other parties: the Environment Agency, Hartlepool Borough Council (HBC; the statuary planning authority), Able UK, MARAD, the US Environmental Protection Agency and the ships themselves (see also Hillier 2006).

Relations certainly varied in their stability. Four ships were towed across the Atlantic, at the behest of MARAD and the Environmental Protection Agency, prior to completion of legal proceedings—even though the UK's regulator had by this time withdrawn Able's licence to scrap the vessels.[7] On the ships' arrival in Hartlepool, the UK government agreed that the law required the ships to be returned to the USA; a decision challenged by the US authorities on the basis that it would not be "safe, environmentally sound, or practicable to [do so]" due to "weather concerns at this time of year" (Beckett 2003, cited in Bhattacharya 2003). The material presence of the ships in Teesside split the opposition with "some people saying 'look, the ships are already here' and another camp of people saying 'they're here illegally, they should go back' ... it wasn't so clear cut then" (FOE representative). In other words, the ships, and their deteriorating hulls, now weakened the claims of campaigners:

> we've got reports saying that the hulls of the ships were so damaged that when divers went under in the James River to do maintenance, the ships were so barnacled up they couldn't see where leaks were coming ... so [if you] argue with the 'too dangerous to come and make that voyage in those waters', then it's understandable why people might suggest they must be too dangerous to go back" (FOE representative).

As a result of the legal proceedings and the inclement weather conditions, the four ships, key elements of the reconfigured and rescaled network, remained berthed at Graythorp as a ghostly material presence, fixed in their new spatial and temporal moorings, unable to be either dismantled or returned to the USA. On the basis of one missing intermediary (the environmental assessment statement), the entities held together as part of Able's "ship recycling" network began to disassemble. In the words of the then local MP, Peter Mandelson: "what happened was that one loose thread was pulled and the rest of the embroidery of permissions and licences promptly unravelled" [Environment Food and Rural Affairs (EFRA) Committee 2003:4].

Regular press releases—appearing on the FOE website and circulated to local, national and international media—included ample statistics. Among these, references to the volumes of waste contained within two of the ships—"more than 800 tonnes of American toxic waste, with more than 500 tonnes of asbestos and 300 tonnes of solid PCBs set to be buried locally" (FOE 2003b)—gave a material form to the campaign's environmental and social injustice claims. These texts (and the embedded numerical formalisms that simplified the messy physical properties of the ships) were circulated in a constant effort to make the network durable through time and space (Latour 2005; cf Routledge 2008). The press, as mediators of social relations, presented a critical vehicle for transforming, translating, distorting and modifying the words of FOE and projecting them onto a national (and even international) stage. The claims made by FOE, and the scale framing of the justice issue as one of localised environmental harm visited upon Teesside by the actions of the USA, resonated with sufficient numbers of policy actors at the heart of government that the EFRA Committee (2003) held an emergency 1-day session on the "ghost ships issue". The ghost ships had arrived firmly on national political and policy agendas (Hillier 2006:24).

The High Court decisions did not, however, determine whether the ships would or would not be recycled in Britain, and Able still had the option to submit a fresh application for a new waste management licence—requiring an environmental assessment and proper consultation with local communities. To achieve Able's commercial ship recycling ambitions, it became increasingly necessary for the firm to assemble a set of socio-material associations (and, indeed, to enrol many of the allies loosely connected to the FOE/Impact network) around an alternative (scale) framing of the Hartlepool ghost ships. In January 2005, in a bid to allow the work to go ahead, the company submitted the missing environmental impact statement in support of three planning applications and an application to store hazardous substances. HBC went on to refuse permissions in October 2006 (*Northern Echo* 2006), even though Council officers had recommended approval. The framing of the ghost ships by campaigning groups in terms of the locally damaging impacts of national/international injustices succeeded in convincing local councillors—as the following remark by one Council member underlines.

> Councillor Edna Wright said the ill-health legacy of the town's shipbuilding and industrial past was still keenly felt . . . There has been no scaremongering, the facts are there for all to see. This has been long-awaited and is the right decision for the town, which now gives it the go-ahead to become a tourist destination rather than an international toxic tip (*Northern Echo* 2006).

Able UK developed and deployed what Kurtz has described as counter scale frames (Kurtz 2003:907), drawing on much the same repertoire of scale idioms but fundamentally redefining the problem. At the scale of the ships themselves, the company was quick to refute claims about the material constitution (ie volumes of hazardous waste) and integrity of the ships, absencing or subduing the toxic presence FOE had so powerfully enrolled, countering that "The ships are not heavily contaminated with toxic substances" (Able UK 2003a), that "[i]n truth these vessels contain only very small amounts of residual fuel oil in their tanks" (Able UK 2003b) and that "there is no more risk to the marine environment during transportation than for any other ship on the high seas" (Able UK 2003a).

Crucially, the firm claimed that, rather than exacerbating local problems, the ships were part of the solution to local–regional economic decline. Able UK tacitly deployed the poor state of the local and sub-regional economies in an effort to enrol "the north-east" (local MPs, businesses, unions, the media and the wider public) into this framing of the ships as a sign of economic growth and new jobs. The power of this counter discourse stemmed partly from Able's deployment of the claim across a series of territorial scales. The firm argued that their Teesside Environmental Recycling and Reclamation Centre would "[boost] the opportunities for the Tees Valley" (Able UK 2005), "[place] the North East at the forefront of this emerging sector" (Able UK 2006a) and, more colloquially (deploying the identity of a knowledgeable insider) "establish our area as a centre of excellence for marine recycling" (Able UK 2006b). At a finer scale, Able materialised real jobs, people and businesses that would benefit: "over 749 quality permanent jobs and benefits to North East Suppliers of around £50m per annum" (Able UK 2006c), the physical presence of "letters of support from individuals and companies from across the North East region" (Able UK 2006c) and the voices and words of MP's sitting on the EFRA Parliamentary Committee who argued that the facility would provide "valuable business and jobs for Hartlepool and the wider North East" (Able UK 2004).

The scalar politics adopted by Able UK went further in claiming that the facility would benefit "the whole of the UK" (Able UK 2007). In a number of ways, then, the national (or outside) level became important in shaping the emerging trajectory of this controversy. Local MPs, crucial in enrolling (national) political support, became key advocates in a strategy that arguably divided the local community (and press). Ashok Kumar, MP for Middlesbrough South and East Cleveland, for instance, described those who protested against the US ghost ships as a "rag bag of mad men and women who obviously don't appreciate what this work means to our area" (cited in Hartlepool Friends of the Earth Media Group 2004) and used the scale idiom of insider/outsider (Kurtz 2003) to present those opposed to the scheme, whether in the north-east or beyond, as interfering and ill-informed "outsiders" who didn't understand the local area and the concerns of its residents.

The following quote from Frank Cook, MP for Stockton North (in evidence to the EFRA Committee) offers a rather different use of insider/outsider. Cook cites the Environment Editor of the *Independent on Sunday*, Geoffrey Lean, and his criticisms of "a scare that never was" and, in doing so, firmly enrols a key figure in the national press into the Able "ship recycling" network. He specifically highlights Lean's description of the campaign to stop the ships being dismantled at Hartlepool as "one of the more outrageous pieces of spin recently to be inflicted on the British public" and the comparison drawn with "merchant ships ... ending their days in the horrendous breakers yards at Alang in India where they are broken apart in the open air with no safety consideration or environmental protection whatsoever and where literally thousands of workers meet their deaths as a result of handling toxic substances" (Cook 2003). It is a move that also worked to define the arguments of the industry in terms of international distributional injustices (pointing to the more significant environmental and health problems being felt in parts of South Asia) and thus refuting the claims of Impact/FOE. A similar point is made by Peter O'Brien (2006, cited in *Northern Echo* 2008), policy officer with the Northern Trades Union Congress: "For far too long, ships have been dismantled in countries where there has been a complete disregard for workplace health and safety and the local environment". It is a scale frame which linked the four ships berthed at Hartlepool to wider issues of global justice and equity—with much of its power effected by the "presence" of distant workers in Alang (and on other South Asian beaches) subjected to dangerous working conditions and impacted by a weak regulatory environment. O'Brien makes the geographical association (or comparison) with the UK situation and the "robust health and environmental safeguards" which would enable "regions such as the North-East" to develop leading-edge capacity to dismantle and recycle ships. These frames, that knitted together the social and the material, were successful in enrolling, and indeed mobilising, key allies into this reading of the critical scales and sites of environmental injustice. In the words of Able's Chairman and Chief Executive, Peter Stephenson: "we now see key public figures such as local MPs Peter Mandelson, Frank Cook and Middlesbrough['s] Mayor ... being joined by a leading environmental journalist and a local newspaper editor in exposing the Friends of the Earth myth machine and underlining the benefits which our contract can bring to both the local economy and the environment" (Able UK 2003c).

The scale framing of the "ghost ships" in terms of sub-regional (economic) development and global (safety) justice, won key national political allies—convincing members of the EFRA inquiry that "the UK has the potential to establish an industry in ship dismantling which can be done safely and offer economic benefits to the communities in which is it carried out" (Paragraph 54, House of Commons 2004). The government response in 2005, and the subsequent UK Ship Recycling Strategy (DEFRA 2007), endorsed this recommendation and recognised a need for the encouragement and establishment of high-quality ship recycling facilities in the UK (DEFRA 2005, 2007). In 2007 HBC gave conditional approval to revised planning applications from the firm, after fresh government policy meant that the council could no longer block planning permission (BBC News 2007). In the summer of 2008 the Environment Agency issued a waste management licence authorising the keeping, treating or disposal of controlled waste—thus giving Able UK the green light to dismantle the ghost ships (Blackburn 2008).[8]

FOE's involvement with the Impact group came to an end in early 2005, and the latter's campaigning on this and other EJ issues has subsequently stagnated and tailed off—lacking the numbers and organisational capacities to maintain a presence or extend its role beyond essentially parochial local matters. The tensions and ambiguities around the appropriate scale(s) of meaning and regulation—particularly given the involvement of an (inter)national NGO—are important in understanding the shrinkage of this particular EJ network. Some Impact members had raised concerns that the ghost ships issue (an issue that tied in with FOE's national agenda) hadn't been adequately communicated to the group and had been prioritised at the expense of problems important to the local community (such as an incinerator campaign)—again increasing the fragility of the local network (Scandrett 2005), particularly when FOE's involvement declined. That the arrival of the ghost ships occurred early in the development of Impact clearly resulted in some disorientation of the group in the tailwind of the campaign (Scandrett 2005). The overall scale framing of the Teesside project (by FOE) also resonates with Miller's (1994) examination of the (negative) consequences of the decision by the American Nuclear Freeze Campaign of the 1980's to decentralise and localise its activities. That campaign focused its efforts at the local scale, with opponents successfully availing themselves of scale resources drawn from wider, specifically national scales (see also Williams 1999). A review of the Impact project, undertaken by FOE Scotland (Scandrett 2005), similarly highlighted that the predefined, campaign-led objectives had meant that limited attention was paid to developing local capacities or, indeed, extending the network beyond the place-specific conditions of Teesside.

Rescaling Environmental Justice: Mobilising the Region

It is evident from the collective reflection on the part of FOE that the scale framing embedded in the Teesside project, although partly successful in achieving the aims set, had been more limited in promoting a resilient and self-sustaining grassroots network capable of achieving impacts at the local (or wider) scale of regulation. In developing their EJ work in the north-east, key figures within FOE talked increasingly of the need to assemble more diverse (and multi-scale) networks and frames of justice. Since 2005 FOE have directed the articulation of (and action on) EJ towards the UK's emergent regional governance landscape. As part of its state modernisation project of "administrative decentralisation" in 1999, the "New Labour" administration embarked on a path of devolution and regionalisation. This rescaling of the state or "new regionalism" (Amin 1999; Painter 2008) has included the creation of regional development agencies (RDAs) with the aspiration of transforming regional economies, the strengthening of regional government offices and the delivery of accountability through (unelected) regional assemblies.[9] More recently, the government's "sub-national economic development and regeneration review" sets out a number of reforms including handing new powers to RDAs and local authorities (with the abolition of Regional Assemblies in 2010), giving regions a greater say in the distribution of funding (HM Treasury, Department for Business Enterprise and Regulatory Reform, Communities and Local Government 2007) and a lead role for RDAs in effective stakeholder engagement and management. In the words of Voluntary Organisations' Network North East, "The policy hooks and levers to support the [voluntary and community] sector's role and voice could arguably not be greater" (2008:10). It is a position that is echoed by a senior member of FOE:

> Because of the rise of the regional economic development agency, and the power that they [will have] the way that I'm looking at this . . . there is a way of talking about the NE and therefore presenting that challenge back to those bodies to say actually the kind of development, the kind of economic development that you're seeking for the region isn't necessarily solving a lot of your long term and persistent problems . . .

The EJ Networking Group (running 2006–2009), funded by a regional charity, is one of a number of FOE EJ campaign projects that, while building on the Teesside work, seek to effectively jump scales (Smith 1984) in order to capitalise on these new regional (policy) "hooks and levers". The EJ Networking Group aims to provide the resources and capacities to build social capital, assisting communities to work collectively and establishing a more durable EJ network over space and time. The network was aimed at bringing together a "broad cross section of community leaders, policy makers, academics and activists" and, through this mechanism, to "influence decision making processes and affect policy in the North East" (FOE 2007:1). Work in the USA has similarly suggested that coalition building is critical to the breadth, stability and power of EJ networks (eg Pellow 2001). It is a move that responds to the scalar and relational challenges thrown up by the Teesside work and the conclusion that to achieve real impacts diverse tactics were needed in building on alliances between Impact, FOE and other actors (Scandrett 2005). The EJ Networking Group also explicitly sought to mobilise a regional interest, one that prioritises EJ issues, but in the name of the region as a whole (Hudson 2007; Jones and MacLeod 2004) and through political opportunities and leverage at this scale "force this [pressure] back through the [national] system" (FOE representative). Through such scalar tactics[10] key social actors within FOE have positioned the organisation as a critical regional intermediary or node in mediating between and affecting the action of environmental regulators, business and wider civil society (cf Medd and Marvin 2008). It is a regional-scale articulation of EJ which is strategically assembled to capitalise on what, at this point in time, represented the most favourable political and economic opportunity structures (see also Pellow 2001).

Conclusion

In this chapter we have considered how research concerned with the scalar dynamics of EJ might constructively be extended. To do so we have drawn on a small but developing body of work that has examined EJ as a politics of scale. Our work similarly observes that the scaling of injustice is strategically oriented to enrol allies, build relational power and achieve specified political ends—and, as such, is an integral part of strategies of empowerment and disempowerment (also Kurtz 2003; Smith 1984, Swyngedouw 2004). In this chapter we have sought to extend this work by more fully attending to the entities (human and non-human) that act, and are continually (re)negotiated, to "scale" more or less durable assemblages of justice. Bringing to bear insights from ANT we have begun to sketch out some of the ways in which non-humans (eg the internet, texts, industrial pollutants, ships and their constituent elements, ocean systems, local wildlife) are woven together with places (Hartlepool, Teesside, Alang), human bodies, words and political–economic structures in the enactment of environmental (in)justice (cf Bennett 2005). We take as an illustrative case the scalar politics that surrounded the movement of ships from the US "ghost fleet" to Hartlepool in north-east England for dismantling, recycling and hazardous waste management. These ships embodied, simultaneously, chemical and physical properties, socio-economic/political characteristics and cultural and symbolic meanings (Swyngedouw 2006a:5). Our account scrutinises the scalar discourses, tactics, actions, representations and so forth—that lead to network expansion, convergence, bifurcation or dissolution (Routledge, Cumbers and Nativel 2007). We believe that this approach does open up some important critical avenues and challenges for future EJ research.

Firstly, the ghost ships campaign (led by FOE) centred on a reading of EJ in terms of distributional inequalities (national and international) that impinged on socially and economically deprived communities already impacted by a legacy of environmental damage and degradation. The claims of (local) exposure of animal and human bodies to hazardous chemicals contained within the deteriorating vessels, targeted at the national scale of regulation, ultimately failed to move the issue substantially beyond the parameters of local problems and effects. This can be contrasted with Able UK's (political) success in elevating their justice claims to other (national and global) scales of meaning, stressing in particular the size and severity of inequities in South Asian states, which resonated with strategic (inter)national political priorities. The FOE/Impact campaign underlines that mobilisation defined largely by a scale of meaning (in this case local impacts) which does not clearly link to efficacious scales of regulation may well be limited in its impact and longevity (Howitt 2003; Pellow 2001). Similarly, Pellow's (2001) work in the USA reveals how tactical repertoires of social movements have changed and re-scaled to successfully respond to a new (neoliberal) political–economic landscape in which corporations figure strongly. Our own research points to the ways in which changes in the UK's governance landscape have precipitated a scale reorientation on the part of EJ activists. Here, then, we see scope for work that pursues the "How? Why? and When?" of EJ movement re-scaling tactics—to better understand the processes and practices underpinning these scale politics, and also the drivers and limits to such strategies (cf Holifield 2004; Swyngedouw 2006b)—particularly where coalitions are built between actor groups operating at a number of geographical scales.

In taking this challenge forward, we might productively explore other spaces of justice—with respect to different national (political–economic and cultural) contexts and alternative scales of mobilisation. On the former point, there is much to be gained from extending this style of analysis to consider the translation of *environmental justice* which, although travelling well beyond the US (for instance Debbané and Keil 2004; Walker and Bulkeley 2006), has remained a problematic and heavily contested concept.[11] To develop this a little further, following the assemblage of EJ networks elsewhere—such as the strong regional associations in the USA—would enrich our understanding of the scalar dynamics, the political economy and democratic efficacy of EJ activism.

We have begun to explore the role of non-human actors implicated in shaping the course of EJ activism and in enabling social actions and relations to persist. These actors all played their part in the unfolding scalar politics of the ghost ships saga. Castree (2002:135) contends that while it may well be liberating to reveal these myriad non-human actants, it will count for little if they are merely described in their subjugation to others (ie human actors). Certainly, the artefacts and entities we have described here do impinge on the world around them, but the role of social actors in influencing these non-human others often remains. The issue of power therefore remains critical. How we work with ANT approaches in EJ research—that is, the "shared capacities" to which Castree (2002) refers alongside inevitable inequalities in agency, with action directed by some more than others—poses some real challenges for future engagements. Following up this point, the scaling of EJ touches upon important questions about the socio-material assemblage of new forms of political solidarity such as we now see occurring around *transboundary* and *transcalar* issues like climate change (Agyeman, Bulkeley and Nochur 2007) while, simultaneously, other activists (see Rixecker and Tipene Matua 2003) are working on issues such as bioprospecting/biopiracy at the *subcellular* or genetic scale. EJ activisms at these scalar extremes raise critical and as yet unanswered questions about the distribution of power across networks between social and natural entities.

Finally, we have touched on the role of different kinds of intermediaries (whether IT systems, a community worker, the media or FOE itself) which connect actors, and, in doing so, play a role in ordering and defining relationships and, through this, the scaling of EJ. Here we see an important role for research which examines how particular intermediaries are constituted and the functions and processes they perform. Medd and Marvin (2008), for instance, have analysed the efforts of organisations involved with sustainable water management to translate national strategies into local practices across multiple spatialities (see also Routledge 2008). In a similar way, FOE emerges as an intermediary of EJ—particularly in the strategic repositioning of the organisation to broaden links with and between (regional) actors and networks. The (changing) nature, work and agency of different intermediaries remains an underdeveloped arena for EJ researchers. It is only through extending these scales of inquiry that we can, in any meaningful sense, describe, conceptualise and indeed respond to the politics of environmental (in)justice.

Acknowledgements

Thanks to members of Friends of the Earth England, Wales and Northern Ireland for generously giving their time, and to two anonymous reviewers for their helpful comments. This chapter was, in part, supported by a grant from the Economic and

Social Research Council (RES 000-23-0007). The map was provided by the North East Regional Information Partnership. The usual disclaimers apply.

Endnotes

[1] Williams (1999:60–61) uses the term "first wave" to characterise research that highlighted the broad scope of disproportionate burdens faced by communities of colour and the poor (eg Bryant and Mohai 1992, Bullard 1983, Pollock and Vittas 1995). Collectively, this work places emphasis on the correlations found in the data, with little attention paid to causation.

[2] A characterisation that tends to focus on formal definitions of deprived areas, and the people who live within them, as disadvantaged relative to other more prosperous areas. In reality, there are no simple cut-off points between deprived and other areas; deprivation is a spectrum of social, political and economic processes that are constituted in and through particular places (North et al 2007: 18).

[3] Teesside—the contiguous built-up area stretching from Stockton through Middlesbrough to Redcar—has a population of almost 400,000. It was the name of a county borough between 1968 and 1974, after which it was absorbed into the larger county of Cleveland. The Tees Valley "city region" was created by the local government reorganisation of the county of Cleveland in 1996 (Tees Valley City Region Development Programme 2006)—aimed at reflecting an economic geography based on real flows of people and activity that take place in central urban areas and their environs. The Programme is based around the five towns of Darlington, Hartlepool, Middlesbrough, Stockton on Tees and Redcar with a population of 875,000 (Tees Valley City Region Development Programme 2006).

[4] Much of this analysis draws on interviews and conversations with members of FOE over 2007, following the organisation's developing work on environmental justice in the north-east and beyond.

[5] The Impact discussion group "went from something like 25 members occasionally sending an email to something like 750 emails in one week when the ghost ships came on. It just went berserk ... Some of the emails were from America as well" (FOE representative).

[6] A local FOE group in Hartlepool was established during the ghost ships campaign.

[7] MARAD, which had jurisdiction over the ships, stated: "As we work toward a resolution of these issues between the UK Environment Agency and Able UK, the ships will continue to transit the Atlantic" and contended that "Prior to the ships' departure, MARAD sought and received official approvals from the UK Environment Agency, the UK Maritime Coastguard Agency, the US Environmental Protection Agency and the US Coast Guard" (Environment News Service 2003).

[8] The contract to export the nine remaining ex-naval vessels from the James River in Virginia to Teesside was cancelled in 2007. It is expected that the nine ships will be put up to bid for domestic ship recyclers (Vidal 2007).

[9] It was the government's intention to move towards a degree of self-government for the English regions (Cabinet Office and DTLR 2002), but the rejection by the north-east electorate of proposals for an elected regional assembly has led to no further progress being made on plans for a regional government in England (North, Syrett and Etherington 2007:29). The regional tier's effectiveness has thus been limited by its lack of agenda-setting powers and legitimacy with other stakeholders (North, Syrett and Etherington 2007).

[10] For instance, in 2005 FOE held a joint 'Environmental justice regional policy seminar" with IPPR North (Institute for Public Policy Research)—a regional (and UK) think-tank. The conference represented a space for constituting, performing and

representing the network (Routledge 2008) and for capitalising on the new territorial opportunities presented by regionalism (Somerville 2004).
[11] Here academic debate has centred on philosophical matters—that is, the nature of EJ (justice as distributional equality of environmental risks, as "recognition" or as participation in decision-making processes) and the plurality or universality of justice principles—as well as on epistemological matters, the appropriate scales of investigation and methods of analysis (Walker and Bulkeley 2006).

References

Able UK (2003a) Able hits back at prophets of doom. Press release 3 September. http://www.ableuk.com/ableshiprecycling/press-able-030903.htm Accessed 9 April 2009

Able UK (2003b) Company backs "hypocrisy" attack on ships critics. Press release 22 October. http://www.ableuk.com/ableshiprecycling/press-able-031022.htm Accessed 9 April 2009

Able UK (2003c) Support messages "expose the fear campaign" says company. Press release 17 November. http://www.ableuk.com/ableshiprecycling/press-able-031117.htm Accessed 9 April 2009

Able UK (2004) Able welcomes vindication in MPs ship dismantling report. Press release 11 November. http://www.ableuk.com/pressreleases.shtml Accessed 9 April 2009

Able UK (2005) Licence application—important step forward for TERRC plans. Press release 10 March. http://www.ableuk.com/pressreleases.shtml Accessed 9 April 2009

Able UK (2006a) Update on TERRC planning permission and waste management licence. Press release 8 June. http://www.ableuk.com/pressreleases.shtml Accessed 9 April 2009

Able UK (2006b) Able pledges to appeal over TERRC plans refusal. Press release 12 October. http://www.ableuk.com/pressreleases.shtml Accessed 9 April 2009

Able UK (2006c) Able disappointment as council refuses to think again Press release 6 November. http://www.ableuk.com/pressreleases.shtml Accessed 9 April 2009

Able UK (2007) Disappointment over ships contract—but appeal continues. Press release 30 May. http://www.ableuk.com/pressreleases.shtml (last accessed 1 July 2009)

Agyeman J (2002) Constructing environmental (in)justice: Transatlantic tales. *Environmental Politics* 11:31–53

Agyeman J (2005) *Sustainable Communities and the Challenge of Environmental Justice.* New York: New York University Press

Agyeman J, Bulkeley H and Nochur A (2007) Just climate: Towards a reconstruction of climate activism? In J Isham and S Waage (eds) *Ignition: What You Can Do to Fight Global Warming and Spark a Movement* (pp 135–147). Washington DC: Island Press

Agyeman J, Bullard R and Evans R (2003) *Just Sustainabilities: Development in an Unequal World.* Cambridge, MA: MIT Press

Agyeman J and Evans B (2004) Just sustainability: The emerging discourse of environmental justice in Britain? *Geographical Journal* 2:155–164

Amin A (1999) Regions unbound: Towards a new politics of place. *Geografiska Annaler, Series B* 86:33–44

Basel Action Network (BAN) (2003) *Needless Risk: The Bush Administration's Scheme to Export Toxic Waste Ships to Europe.* http://www.ban.org/Library/Needless%20Risk%20Final.pdf Accessed 9 April 2009

BBC News (2004) Britain's forgotten communities, 21 October. http://news.bbc.co.uk/1/
   hi/programmes/newsnight/3757100.stm Accessed 9 April 2009
BBC News (2007) Ghost ships scrap approved, 25 October. http://news.bbc.co.uk/
   1/hi/england/tees/7062714.stm Accessed 9 April 2009
Benford R (2005) The half-life of the environmental justice frame: Innovation, diffusion
   and stagnation. In D N Pellow and R J Brulle (eds) *Power, Justice and the Environment:
   A Critical Appraisal of the Environmental Justice Movement* (pp 37–53). Cambridge,
   MA: MIT Press
Benford R D and Snow A D (2000) Framing processes and social movements: An
   overview and assessment. *Annual Review of Sociology* 26:611–639
Bennett J (2005) The agency of assemblages and the North American blackout. *Public
   Culture* 17:445–465
Beynon H, Hudson R and Sadler D (1994) *A Place Called Teesside.* Edinburgh:
   Edinburgh University Press
Bhattacharya S (2003) Toxic US ghost ships should go home. NewScientist.com news
   service 6 November. http://www.newscientist.com/article/dn4358.html Accessed 9
   April 2009
Blackburn M (2008) Ghost ships go-ahead as licence is issued. *Evening Gazette*
   26 June. http://www.gazettelive.co.uk/news/teesside-news/2008/06/26/ghost-ships-
   go-ahead-as-licence-is-issued-84229-21154543/ Accessed 9 April 2009
Boardman B, Bullock S and McLaren D (1999) *Equity and the Environment.* London:
   Catalyst Trust
Bryant B and Mohai P (1992) *Race and the Incidence of Environmental Hazards.*
   Boulder, CO: Westview Press
Bullard R D (1983) Solid waste sites and the black Houston community. *Sociological
   Inquiry* 53:273–288
Bullard R D and Johnson G S (2000) Environmental justice: Grassroots activism and
   its impact on public policy decision making. *Journal of Social Issues* 56:555–578
Bush J, Moffatt S and Dunn C (2001) "Even the birds round here cough": Stigma, air
   pollution and health in Teesside. *Health and Place* 7:47–56
Cabinet Office and DTLR (2002) *Your Region, Your Choice; Revitalising English
   Regions.* London: The Stationery Office
Callon M (1991) Techno-economic networks and irreversibility. In J Law (ed) *A
   Sociology of Monsters? Essays on power, technology and domination* (pp 132–161).
   London: Routledge
Capek S (1993) The environmental justice frame: A conceptual discussion and
   application. *Social Problems* 40:5–24
Castree N (2002) False antitheses? Marxism, nature and actor-networks. *Antipode*
   34:111–146
Cook F (2003) Memorandum submitted by Frank Cook, MP for Stockton North.
   In House of Commons Environment, Food and Rural Affairs Committee (EFRA
   Committee), *US Ghost Ships*, Minutes of Evidence & Memoranda, HC 1336,
   Session 2002–2003. http://www.publications.parliament.uk/pa/cm200203/cmselect/
   cmenvfru/1336/311915.htm (last accessed 1 July 2009)
Debbané A and Keil R (2004) Multiple disconnections: Environmental justice and urban
   water in Canada and South Africa. *Space and Polity* 8:209–225
Delaney D and Leitner H (1997) The political construction of scale. *Political Geography*
   16:93–97
Department for Communities and Local Government and Tees Valley Strategy Unit
   (2006) Tees Valley City Region Business Case and City Region Development
   Programme. http://www.egenda.stockton.gov.uk/aksstockton/images/att1022.DOC
   Accessed 9 April 2009

DEFRA (2005) In House of Commons Environment, Food & Rural Affairs Committee, *Dismantling Defunct Ships in the UK: Government Reply to the Committee's Report, Second Report of Session 2004–2005*. Report, together with formal minutes (pp 5–15). London: The Stationery Office

DEFRA (2007) *UK Ship Recycling Strategy*. London: The Stationery Office

DEFRA (2008) *Aarhus Convention Implementation Report*. London: The Stationery Office

Dobson A (1998) *Justice and the Environment: Conceptions of Environmental Sustainability and Dimensions of Social Justice*. Oxford: Oxford University Press

Environment Agency (2004) *Addressing Environmental Inequalities*. http://www.environment-agency.gov.uk/commondata/105385/ca221final_888457.pdf Accessed 9 April 2009

Environment, Food and Rural Affairs Committee (EFRA Committee) (2003) *US Ghost Ships*, Minutes of Evidence & memoranda, HC 1336, Session 2002–2003. http://www.publications.parliament.uk/pa/cm200203/cmselect/cmenvfru/1336/3111901.htm (last accessed 1 July 2009)

Environment News Service (2003) British agency withdraws permission to scrap ghost ships, 31 October. http://www.ens-newswire.com/ens/oct2003/2003-10-31-04.asp Accessed 9 April 2009

Every Action Counts (2008) Living in the shadow of industry—Impact, Teesside. http://www.everyactioncounts.org.uk/en/fe/page.asp?n1=230&n2=6&n3=491&n4=449 Accessed 9 April 2009

Faber D (2005) Building a transnational environmental justice movement: Obstacles and opportunities in the age of globalization. In J Bandy and J Smith (eds) *Coalitions Across Borders: Negotiating Difference and Unity in Transnational Struggles Against Neoliberalism* (pp 43–68). New York: Roman & Littlefield

FOE (2001) *Pollution and Poverty—Breaking the Link*. London: Friends of the Earth. http://www.FOE.co.uk/resource/briefings/pollution_and_poverty.pdf Accessed 9 April 2009

FOE (2003a) D Day for ghost ships. Press release 10 September. http://www.FOE.co.uk/resource/press_releases/d_day_for_ghost_ships.html Accessed 9 April 2009

FOE (2003b) USA sets precedent for dumping its waste overseas. Press release 18 November. http://www.FOE.co.uk/resource/press_releases/usa_sets_precedent_for_dum.html Accessed 9 April 2009

FOE (2003c) Ghost ships in court battle. Press release 8 December. http://www.FOE.co.uk/resource/press_releases/ghost_ships_in_court_battl.html Accessed 9 April 2009

FOE (2003d) Ghost ships—fact not fiction. Press release 8 November. http://www.FOE.co.uk/resource/press_releases/ghost_ships_fact_not_ficti0.html Accessed 9 April 2009

FOE (2004) Minister jumps gun on the ghost ships disposal. Press release 22 April. http://www.FOE.co.uk/resource/press_releases/minister_jumps_gun_on_ghos_22042004.html Accessed 9 April 2009

FOE (2007) Northern Rock Environmental Justice Networking Group. Internal briefing document

Freudenberg W (1997) Contamination, corrosion and the social order: An overview. *Current Sociology* 45:19–39

Guardian (2004) Action call to control ghost ships, 12 May. http://www.guardian.co.uk/print/0,,4921923-110826,00.html Accessed 9 April 2009

Gilligan E and Zagrovic C (2004) Joining up campaigning and community development in Teesside. *CONCEPT* 14:7–11

Hartlepool Friends of the Earth Media Group (2004) In House of Commons Environment, Food and Rural Affairs Committee (EFRA Committee) *Dismantling Defunct Ships in the UK, Eighteenth Report*. HC 834, Session 2003–4. http://www.publications.parliament.uk/pa/cm200304/cmselect/cmenvfru/834/834we10.htm (last accessed 1 July 2009)

Health Protection Agency (2005) *Health Protection in the 21st Century—Understanding the Burden of Disease; Preparing for the Future*. London: Health Protection Agency

Heynen N C (2003) The scalar production of injustice within the urban forest. *Antipode* 35:980–998

Hillier J (2006) Assemblages of justice? From the ghost ships of Graythorp to a UK ship recycling strategy. Paper for Conference on Democratic Network Governance, Roskilde, 2–3 November. http://www.ruc.dk/upload/application/pdf/8b5657af/Assemblages%20of%20Justice,%20From%20the%20Ghost%20Ships%20of%20Graythorp%20to%20a%20UK%20Ship%20Recycling%20Strategy,%20by%20Jean%20Hillier.pdf Accessed 9 April 2009

HM Treasury, Department for Business Enterprise and Regulatory Reform, Communities and Local Government (2007) *Review of Sub-national Economic Development and Regeneration*. London: The Stationery Office

Holifield R (2004) Neoliberalism and environmental justice in the United States Environmental Protection Agency: Translating policy into managerial practice in hazardous waste remediation. *Geoforum* 35:285–297

House of Commons Environment, Food & Rural Affairs Committee (House of Commons) (2004) *Eighteenth Report of Session 2003–2004*. Report, together with formal minutes, oral and written evidence. London: The Stationery Office

Howitt R (1993) A world in a grain of sand: Towards a reconceptualization of geographical scale. *Australian Geographer* 24:33–45

Hudson R (1998) Restructuring region and state: The case of northeast England. *Tijdschrift voor Economische en Sociale Geografie* 89:15–30

Hudson R (2006) Regional devolution and regional economic success: Myths and illusions about power. *Geografiska Annaler, Series B-Human Geography* 88B:159–171

Hudson R (2007) Regions and regional uneven development forever? Some reflective comments upon theory and practice. *Regional Studies* 41:1149–1160

Impact (2003) *The "Impact" Constitution* 9 January. http://www.impact-teesside.org/i_const.htm Accessed 9 April 2009

Jones M and MacLeod G (2004) Regional spaces, spaces of regionalism: Territory insurgent politics and the English question. *Transactions of the Institute of British Geographers* NS 29:433–452

Kroll-Smith J and Couch S (1990) *The Real Disaster is Above Ground*. Lexington, KY: University Press of Kentucky

Kurtz H (2003) Scale frames and counter-scale frames: Constructing the problem of environmental injustice. *Political Geography* 22:887–916

Laurier E and Philo C (1999) X-morphising: A review essay of Bruno Latour's "Aramas or the love of technology". *Environment and Planning A* 31:1047–1071

Latour B (1987) *Science in Action*. Milton Keynes: Open University Press

Latour B (2005) *Reassembling the Social: An Introduction to Actor-Network-Theory*. Oxford: Oxford University Press

London Sustainability Exchange (2004) Environmental justice in London: Linking the equalities and environment policy agendas. http://www.lsx.org.uk/docs/page/2604/Environmental%20Justice%20in%20London%20-%20Linking%20the%20Equalities%20and%20Environment%20Policy%20Agendas.pdf (last accessed 1 July 2009)

MacLeod G (1999) Space, scale and state strategy: Rethinking urban and regional governance. *Progress in Human Geography* 23:503–527

McLaren D, Cottray O, Taylor M, Pipes S and Bullock S (1999) The geographic relation between household income and polluting factories. http://www.FOE.co.uk/resource/reports/income_pollution.html#N_6_ Accessed 9 April 2009

Medd W and Marvin S (2008) Making water work: Intermediating between regional strategy and local practice. *Environment and Planning D: Society and Space* 26:280–299

Miller B (1994) Political empowerment, local–state relations, and geographically shifting political opportunity structures: Strategies of the Cambridge, Massachusetts peace movement. *Political Geography* 13:393–406

Murdoch J (1997) Towards a geography of heterogeneous associations. *Progress in Human Geography* 21:321–337

Nicholls W J and Beaumont J R (2004) The urbanisation of justice movements? Possibilities and constraints for the city as a space of contentious struggle. *Space and Polity* 8:119–135

North D, Syrett S and Etherington D (2007) *Devolution and Regional Governance: Tackling the Economic Needs of Deprived Areas.* York: Joseph Rowntree Foundation

Northern Echo (2006) Appeal pledge after ghost ships decision, 13 October. http://archive.thenorthernecho.co.uk/2006/10/13/232000.html Accessed 9 April 2009

Northern Echo (2008) North's ghost ships scrap industry wins backing, 28 June. http://www.thenorthernecho.co.uk/news/indepth/ghostships/display.var.809684.0.norths_ghost_ships_scrap_industry_wins_backing.php Accessed 9 April 2009

O'Brien P, Pike A and Tomaney J (2004) Devolution, the governance of regional development and the Trade Union Congress in the North East region of England. *Geoforum* 35:59–68

Paasi A (1991) Deconstructing regions: Notes on the sales of spatial life. *Environment and Planning A* 23:239–256

Paasi A (2004) Place and region: Looking through the prism of scale. *Progress in Human Geography* 28:536–546

Painter J (2008) Cartographic anxiety and the search for regionality. *Environment and Planning A* 40:342–361

Pellow D N (2001) Environmental justice and the political process: Movements, corporations, and the state. *The Sociological Quarterly* 42:47–67

Pellow D N and Brulle R J (2005) *Power, Justice, and the Environment.* Cambridge, MA: MIT Press

Phillimore P and Moffatt S (1999) Narratives of insecurity in Teesside: Environmental politics and health risks. In J Vail and J Wheelock (eds) *Fostering Insecurity: Policies and Lived Experiences* (pp 137–153). London: Routledge

Pollock P H and Vittas M E (1995) Who bears the burdens of environmental pollution? Race, ethnicity and environmental equity in Florida. *Social Science Quarterly* 76:294–310

Pulido L (1996) A critical review of the methodology of environmental racism research. *Antipode* 28:142–159

Rixecker S S and Tipene-Matua B (2003) Maori Kaupapa and the inseparability of social and environmental justice: An analysis of bioprospecting and a people's resistance to (bio)cultural assimilation. In J Agyeman, R D Bullard and B Evans (eds) *Just Sustainabilities: Development in an Unequal World* (pp 252–268). Cambridge, MA: MIT Press

Robinson F (2002) The North East: A journey through time. *City* 6:317–334.

Routledge P (2008) Acting in the network: ANT and the politics of generating associations. *Environment and Planning D: Society and Space* 26:199–217.

Routledge P, Cumbers A and Nativel C (2007) Grassrooting network imaginaries: Relationality, power, and mutual solidarity in global justice networks. *Environment and Planning A* 39:2575–2592

Rutland T and Aylett A (2008) The work of policy: Actor networks, governmentality, and local action on climate change in Portland, Oregon. *Environment and Planning D: Society and Space* 26:627–646

Sandweiss S (1998). The social construction of environmental justice. In D Camacho (ed) *Environmental Injustices, Political Struggles* (pp 31–58). Durham, NC: Duke University Press

Scandrett E (2005) *Teesside Review Report*. Friends of the Earth. Internal report

Smith N (1984) *Uneven Development*. Oxford: Blackwell

Somerville P (2004) State rescaling and democratic transformation. *Space and Polity* 8:137–156

Stephens C, Bullock S and Scott S (2001) Environmental justice: Rights and means to a healthy environment for all. Special Briefing Paper 7, ESRC Global Environmental Change Programme

Swyngedouw E (2004) Globalisation or "glocalisation"? Networks, territories and rescaling. *Cambridge Review of International Affairs* 17:25–48

Swyngedouw E (2006a) United Nations Development Programme background paper: Power, water and money: Exploring the nexus. http://hdr.undp.org/en/reports/global/hdr2006/papers/Swyngedouw.pdf Accessed 9 April 2009

Swyngedouw E (2006b) "Impossible/undesirable sustainability and the post-political condition." Paper presented at the Royal Geographical Society with the Institute of British Geographers Conference, London, September

Taylor D (2000) The rise of the environmental justice paradigm: Injustice framing and the social construction of environmental discourses. *American Behavioral Scientist* 43:508–580

Tees Valley City Region Development Programme (2006) *The Tees Valley City Region Business Case*. http://www.teesvalley-jsu.gov.uk/tvcr/reports/jsureport.pdf Accessed 9 April 2009

Tees Valley Joint Strategy Unit (2000) *The Tees Valley Economy* Issue 54, September. http://www.teesvalley-jsu.gov.uk/old/jsu5_tve.htm Accessed 9 April 2009

Towers G (2000) Applying the political geography of scale: Grassroots strategies and environmental justice. *Professional Geographer* 52:23–36

Vidal J (2003) Town turns its back on America's toxic ships. *The Guardian* 14 October. http://www.guardian.co.uk/science/2003/oct/14/sciencenews.theguardian lifesupplement Accessed 9 April 2009

Vidal J (2007) British firm loses contract to dismantle nine US "ghost ships". *The Guardian* 31 May. http://www.guardian.co.uk/environment/2007/may/31/pollution.uknews Accessed 9 April 2009

Voluntary Organisations' Network North East (2008) *The VINE: Focus on Regional Governance*, Issue 27

Walker G and Bulkeley H (2006) Geographies of environmental justice. *Geoforum* 37:655–659

Walker G, Mitchell G, Fairburn J and Smith G (2003) Environmental quality and social deprivation. Phase II: National Analysis of Flood Hazard, IPC Industries and Air Quality. R&D Project Record E2-067/1/PR1. Bristol: The Environment Agency

Walker G P, Mitchell G, Fairburn J and Smith G (2005) Industrial pollution and social deprivation: Evidence and complexity in evaluating and responding to environmental inequality. *Local Environment* 10:361–377

Williams F (2003) Campaigners protest as US Nay's toxic fleet heads across the Atlantic. *Financial Times* 23 October. http://www.ban.org/ban_news/campaigners_protest_us_us.html Accessed 9 April 2009

Williams R (1999) Environmental injustice in America and its politics of scale. *Political Geography* 18:49–73

Chapter 8

Defining and Contesting Environmental Justice: Socio-natures and the Politics of Scale in the Delta

Julie Sze, Jonathan London, Fraser Shilling,
Gerardo Gambirazzio, Trina Filan
and Mary Cadenasso

Introduction

This chapter examines a contemporary policymaking process known as the "Delta Vision", intended to "identify a strategy for managing the Sacramento–San Joaquin Delta as a sustainable ecosystem that would continue to support environmental and economic functions that are critical to the people of California".[1] The Delta, located in the Central Valley Region in California, is a complex aquatic and terrestrial wetland ecosystem, through which approximately 40 million acre-feet[2] of water flow each year. Eighteen million acre-feet of this water are diverted for agriculture and urban consumptive use, including diversion to southern CA (Lund et al 2007). The Delta is the largest freshwater (formerly tidal) estuary on the West Coast, encompassing 738,000 acres of wetlands and islands, and 700 miles of meandering waterways (Delta Protection Commission 2007). It is formed by the convergence of the Sacramento and San Joaquin Rivers and the San Francisco Bay. The Delta region bisects California's Central Valley, dividing it into the Sacramento Valley to the north and the San Joaquin Valley to the south. The Central Valley is a vast plain surrounded by mountains and covering nearly 15 million acres—an area as large as England (Johnson, Haslam and Dawson 1993). The Delta is the conveyance mechanism that waters California's massive and global agribusiness and that lubricates the state's growth machine of urban and suburban development. As such, Delta water supplies drinking water to 20 million people from the San Francisco Bay Area to the Los Angeles Basin, nearly half of California's population.

The Delta is considered in "crisis" by politicians, scientists and environmentalists due to a combination of water management decisions, impaired water quality, invasive species, and upstream and upland land-use impacts. Pressures on the Delta include rapid urban and suburban encroachment, agricultural chemical runoff, contamination by mercury and other legacies of the state's mining and industrial production, subsidence of peat soils due to cultivation, and broader-scale, complex climatic changes such as sea-level rise. Together these forces have brought the Delta to the brink of an ecological collapse with the loss of wetlands, negatively impacted water quality and the near-total collapse of several animal and plant populations, including several endangered fish species (Little Hoover Commission 2005). In what otherwise might simply be a vast and peaty wetland, the needs and desires of the state (principally continued economic growth from urban and agricultural development) have collided disastrously with the health and continued function of the ecosystem.

These environmental, land-use and population pressures have prompted a policy response, the Delta Vision. The Delta Vision is a technocratic-managerial process that has further centralized decision-making about the fate of the Delta into the hands of public regulatory agencies and powerful water, utility and agricultural interests. The Delta is the narrative, symbolic, and material site of struggle between the forces of state, capital, and nature. However, in this clash of the titans, the voice, interests, and visibility of human communities, especially socially vulnerable populations, whether defined by race or class, have been marginalized in policy debates on the Delta. This marginalization is neither new, nor surprising, yet it is still important to document and understand. During the long and convoluted history of efforts to guide and shape the workings and fate of the Delta, its ecosystems and its ecological processes, political and economic interests have and continue to circumscribe, discipline, and name the Delta as the unruly place that once was many places and no place at once. The Delta has been shaped by desires to know and to own it through twin projects of furthering capital-intensive economic development and of state-building in California.

Our research focuses on competing definitions and representations of environmental justice and scale in the Delta. Specifically, our analysis is focused on identifying whether and how environmental justice as a conceptual category is understood and interpreted in the Delta Vision process. We seek to highlight what environmental justice in the Delta means, but, to do so, we first articulate several contested questions related to this domain of environmental justice, scale and the Delta. How is the Delta bounded and understood as a both a material place and as a locus of state intervention? What are the spatial boundaries of the Delta? How were these boundaries set and to serve whose interests? What is within its boundaries and what lies outside, and what are the relationships between these inside and outside realms? In sum, how is the conflict over scale at the core of the politics in the efforts to both "save" and "manage" the Delta through the Delta Vision process?

This scalar ambiguity does not merely imply a question of descriptive clarity. Rather, it signifies and provokes fundamental questions of political power. Within the initial act of setting the boundaries of the Delta as a unit of analysis and enacted through policy intervention inheres its "always already political" character. Within the inscription of the Delta's boundaries are complex politics of representation, both of membership and of discourse. What actors are considered legitimate (and illegitimate) stakeholders in the management of the Delta? What issues are legible (and illegible) in the Delta Vision policy framework? More broadly still, the scaling of the Delta sets the terms of the subsequent struggles. This chapter explores both the struggles by contending actors within the Delta as well as the struggles over scale as different political actors seek to bound the issue in a fashion that privileges their interests over others. For the purposes of our analysis, we define the Delta as a geographical region characterized not solely by its political boundaries nor lines drawn on a map, but rather by its environmental history as a site constructed by large-scale human intervention. The Delta is a region that resource extraction (namely water), capital accumulation, human exploitation and engineering have inscribed as a socio-natural space, susceptible to and at the same time driving the ever-present and always-evolving desires of numerous and varied human populations that seek to leave their mark on its particular geographic and physical environment. We draw on theoretical definitions of socio-nature to describe the interrelationships "between society and nature" and socio-ecological products (Swyngedouw 1996) and which are created through the "social production of nature" (Harvey 1996; Smith 1984).

Our methods are based on historical geographic and discursive analyses of the political and social processes by which the socio-nature of the Delta has been constructed and continues to be shaped. This chapter also draws on 12 months of fieldwork examining the Delta Vision Process and over 20 semi-structured interviews with key actors (scientists, policymakers, and activists). Our linking of the historical processes of exploitation of the Delta, especially relating to water, with the contemporary policy process is an explicit rejoinder to decision-makers who actively seek to erase the history of this exploitation. Our analysis is filtered through our multidisciplinary lenses drawn from geography, sociology, ecological science, and the environmental humanities. This chapter is part of a larger research project that examines science and policymaking in the Delta.[3] Our analysis is also informed by participant observation, based upon attendance at, observation of, and participation in a number of Delta Vision Blue Ribbon Task Force meetings and Stakeholder Coordination Group meetings sponsored by the Delta Vision program, as well as several public education meetings sponsored by the Delta Vision program and other organizations.[4] We also draw heavily from environmental justice activist conceptions of the Delta, primarily from an organization known as the Environmental Justice Coalition for Water (EJCW).

One key finding from our data is that "environmental justice", as a meaningful analytic or political dimension, is marginalized within the Delta Vision process, understood as a "special interest", rather than a term that has particular legal/regulatory meaning and activist articulation. The Delta Vision process can be characterized as one in which dominant actors rationalize their positions, in the service of an abstract and ostensibly apolitical capital and a self-legitimizing state. We argue that the production of environmental injustices in the Delta (some of, but not all of which are experienced in racially disproportionate terms) is exacerbated by connected conflicts between knowledge and power, particularly over the scale at which "environmental justice" and the "Delta" are experienced, understood, or acted upon through public policy. Whereas the central vision of the Delta Vision process imagines what it terms "co-equal" goals that balance economy and the environment, our analysis rejects the separation of these two domains. Indeed, our critique of the Delta Vision process foregrounds the embeddedness of the economic/environmental drawing from socio- and techno-natures[5] literatures that highlight the fusion of the social and the natural, particularly through the mechanisms of what Kaika (2005) calls modernity's Promethean project symbolized by water infrastructure like dams enabled through the confluence of engineering, science and modern state formation (Carroll 2006).

As a complex socio-nature located in the geographic, ecological and economic heart of California, the Delta is an appropriate site to develop critical perspectives in environmental justice research that are simultaneously focused on contemporary politics and grounded in the insights from research on environmental history and radical geography. The Delta's socio-nature is shaped by complex and hierarchical social, political, economic, and ecological forces that have thrust the regulatory state and capital resources into action. The Delta's socio-nature is produced by pumping water through and out of the region using a vast network of gargantuan pumps, aqueducts and canals that convey it over hundreds of miles from the wet and less-populated north to the dry and booming Central Valley and southern California. The Delta is also criss-crossed by a vast levee system to "reclaim" and protect riparian lands for settlement. This system is now considered to be on the brink of collapse, as urbanized regions in the Delta face flooding risk on par with post-Hurricane Katrina New Orleans.[6]

At the same time, the Delta is an interconnected system of everyday places where diverse populations live, work, and play. Like California and its Central Valley in general, the Delta embodies extremes of affluence and poverty and faces a myriad of social and environmental problems, including poor air quality, water contamination, and pesticide poisoning (Harrison 2006). Scholars such as geographer Ruth Wilson Gilmore (2007) document the history of how water politics and the rise of agribusiness shaped the political and social landscape in the Central Valley as a site of both extreme wealth generation and conditions of abject poverty. Historical patterns of labor exploitation continue with farm workers (now primarily, but not exclusively Mexican and Mexican–American) toiling in the Central Valley to wrest "nature's bounty" from the soil. Their labor is repaid by low-wage, hazardous and mostly seasonal work, living in relatively low-quality and high-cost housing, with uneven access to safe drinking water as well as educational, health and other services (McWilliams 1999 [1935]; Villarejo et al 2000). Gilmore (2007) links this political and economic history to the region's current status as the state's dumping ground for prisons and environmental pollution.

We argue that the rejection of both environmental justice and technonatural perspectives under Delta Vision represents how contemporary policy processes are recreating and reenacting conflicting structures of knowledge and power that have shaped the Delta and placed it on a path to ecological collapse and which have defined the region by high levels of social and racial injustice over the past 150 years. Through continued state agency and industrial interventions—modernizing engineering and institutional regulation—the Delta extends beyond its current political boundaries and continues to engage our collective geographical and political imaginaries.

The chapter begins by reviewing the literature on environmental justice and scale and water and power. We then highlight the historical formation of the Delta's "socio-nature" (White and Wilbert 2009). The policy context of the Delta Vision process is outlined, as well as its immediate predecessor, California Federal Bay-Delta Program (CALFED), which was an attempt to manage the Delta under a joint state–federal policy process. In contrast to the state-sanctioned perspective on environmental justice, we present the competing social movement articulation of environmental justice in the Delta, primarily through a non-governmental organization, Environmental Justice Coalition for Water. We then offer a conceptual framework for understanding and articulating environmental justice in the Delta that both builds upon environmental justice activist articulations and goes beyond them, through an analysis drawing heavily on the concept of socio-nature. Our analysis of environmental justice in the Delta thus includes the particular problems of where disproportionate environmental pollution occurs along racial lines, and connects these to other pollution problems associated with the Delta. Our broad view of environmental justice draws upon a wide array of critical, historical and theoretical frameworks, from both the ecological and social sciences.[7] The final section considers the implications of our case study as a rejoinder to influential critiques of environmental justice from radical geography, specifically Harvey's (1996) critique of environmental justice movements as militant particularism, and Sywngedouw and Heynen's (2003) critique of environmental justice research for privileging local notions of distributive justice while missing important critical analysis of capitalism that cannot be gained except at broader scales and framed through Marxist urban ecology.

Scaling Environmental Justice and Water in the Delta

The relevance of scale within environmental justice movements has been addressed in a number of important works, emerging from political and radical geographers. Scholars note that: "the continuous reorganization of spatial scales is an integral part of social strategies to combat and defend control over limited resources and/or a struggle for empowerment" (Swyngedouw and Heynen 2003). Struggles over scale are not simply over who controls a given territorial unit, but about the scale at which that unit is defined (Herod 1991, 1997). Thus, we agree with scholars who take scale to be both an empirical and epistemological tool of understanding and representing the world. Scale is not understood as natural but instead both socially produced (derived from social processes, and often social struggle) and socially producing (exerting coercion and hegemony in a Gramscian sense) (Williams 1999:52). Towers (2000), for instance, argues that the environmental justice movement is "defined by scale", or more specifically, by a tension between local scale(s) at which grassroots protest over unwanted pollution takes place, and the broader geographic scales at which the discourse of environmental justice is directed. Williams (1999) focuses on the mismatch between the scales at which the problems of environmental inequality are most clearly manifested and experienced and the scales at which they are produced and can therefore be resolved (or at least targeted for action). Kurtz (2003:891) similarly observes that: "[T]he very concept of environmental injustice precipitates a politics of scale, as the locally experienced problem of burdensome pollution can hardly be resolved at the local scale, whether by capital or the state, when it originates in political and economic relationships that extend well beyond the scale of the locality".

Another scalar problematic is what Kurtz (2003) calls its spatial or scalar "ambiguity". Such ambiguity derives partially from the "modifiable areal unit" problem (Openshaw 1983) in which "different statistical relationships between spatially aggregated data can be derived using different spatial units of analysis" as a result of both "scale effects" of aggregating data at different resolutions and "aggregation effects" of the different groupings of units. The finding of race as a primary factor at certain analytical scales and class (or "normal" market functioning) at other scales is a primary example of this ambiguity. This research on the centrality of scale for environmental justice movements builds upon earlier work by social movement scholars on the articulation of environmental racism and how problem identification contributes to the formation of the environmental justice movement (Sze and London 2008). Whereas the focus of environmental injustice in the academic literature is generally on controversies over specific siting of polluting facilities, our case stands in contrast by focusing on how environmental injustices are geographically and temporally dispersed. This is necessary in part because the nature of water and water politics in the Delta requires defining and categorizing environmental justice as extremely diffuse.

The diffuse property of water and the theoretical and pragmatic implications of this diffuse quality are a central problematic in the geographic literatures on firstly water and secondly power, which collectively highlight the "already always political" character, politics and scale of water. Recent literature on water management in diverse geographies from Australia, Latin America and Europe highlights the ways in which social and economic power flows in and through water and how social conflicts and power inequalities course through river and ecosystem management (Hillman 2006; Swyngedouw 2004). Radical geographers and others ask: What is water for and for whom is it flowing? How is it managed and indeed, "made"? As Kaika (2005:32) argues, the very "nature" of water is highly technological. She writes that "water enters in one end of the [water supply] network as H_2O and subsequently undergoes a chemical and social transformation to end up at the other end (the tap) as potable water, as a commodity properly priced and treated". This literature on water and power parallel the critical conversations about the social construction of environmental justice movements. Thus, the "nature" of water, like that of environmental justice and scale, is an epistemologically contested and charged terrain. Dams and other objects of water "management" systems represent and embody ideologies of progress, modernity and social control (Swyngedouw 1999), as water becomes "controlled, tamed and domesticated" (Kaika 2005:141).

These questions about water and ideology are also connected with scale in that the command over geographic and regulatory scale is itself illustrative of power. Conflicts over the "appropriate scale for organizing water systems (local, watershed, regional, national, transnational) each evoke different power geometries and may lead to radically different socioecological conditions" (Swyngedouw 2004). Competing scale definitions of particular problems associated with water shape how opponents construct their activist campaigns against water-related facilities. One example is in Native activism against a hydroelectric dam that highlighted a historically and culturally based relationship to rivers and water. In the case of the struggle of the Eyeouch (East James Bay Cree) people against the province of Quebec's construction of the LeGrand River hydroelectric complex, Desbiens (2007) charts a "rescaling of water" and a "rescaling of the nation" in which the Eeyouch took their fight across the border to enlist allies in the United States while the Québécois sought to "downscale" the issue as a domestic negotiation. In addition to competing geographic scales are definitions of scale beyond the spatial. Loo argues, in her history of the damming of Peace River in British Columbia, that scales have "sensual, spatial, and temporal dimensions" (2007).

Our case study of environmental justice in the Delta lies at the nexus of these theoretical conversations on environmental justice and scale, and water and power. Our study of the Delta Vision process employs conceptual tools synthesized from the scale and the social movement literatures that Kurtz has termed "scale frames" (2003:894) as well as analyses of scale that are sensual, spatial and temporal. Kurtz describes scale frames as "the discursive practices that construct meaningful (and actionable) linkages between the scale at which a social problem is experienced and the scale(s) at which it could be politically addressed or resolved" (2003:894). The competing nature of scale frames in the Delta is highlighted in conflicts over defining environmental justice issues and in defining the Delta itself. By situating our case study within critical conversations on environmental justice and scale, and water and power, we simultaneously address critiques of the environmental justice movement that take the movement to task for its place-based and "militant particularism" (Harvey 1996:399). In a similar vein, Sywngedouw and Heynen (2003) and Heynen (2003) critique environmental justice scholarship for privileging local notions of distributive justice while missing important critical analysis of capitalism that cannot be gained except at broader geographic and political scales. This critique of environmental justice as local neglects cases in which that scale is consciously chosen for its political efficacy and radical analysis. For instance, Towers (2000:25) counters Harvey's cosmopolitanism to assert that local or "tactical environments may encourage radical goals and visions". Others observe a rejection of the charge of local and reformist identities in examining environmental justice activists as movements struggle to expand both their scale of analysis and networks of solidarity to regional, national and increasingly global systems (Agyeman 2002; Pellow 2007). Likewise, Schlosberg (2007) argues that notions of environmental justice are built upon sophisticated notions and critical analyses of political injustice, often at the scale of the global. Our analysis builds upon this emergent scholarship on environmental justice movements that are attuned to these critiques of environmental justice as primarily a local and reformist frame that pays insufficient attention to the forces of capital.

Historical and Socio-Natural Formations in the Delta: 1850–1980

For thousands of years, the Delta was a place where Native American peoples made their living from the land and waters of the landscape, which also served as the source and center of cultural identity practices for these groups. A mix of freshwater rivers and saline tidal waters created a huge area of tidal marshlands, low-lying islands, floodplains, and wildlands (Thompson 1957; Wolff 2003). The Delta's large indigenous population (primarily, but not exclusively Miwok) used these resources to form a thriving material culture, but dwindled rapidly in the mid-1800s with the introduction of European explorers and their associated diseases and violent expropriation of land. The region quickly became populated with settlers and others seeking to exploit its natural resources like beaver and mink (Nash 2006). Then gold was found in a Delta tributary, the American River, in 1848, and the urgent, irrevocable re-imagination and transformation of the region and the state began.

After 1860, the spatial imaginary of the Delta landscape was disrupted and consolidated as water policy and large-scale engineering and irrigation projects helped to "win" the West and California in particular in the nineteenth century (Hundley 2001; Pisani 1984; Worster 1985). People in distant cities claimed ownership of the land and became wealthy from the productive labor of people who toiled on the land but could never own it, due to racial exclusions on land ownership targeting, at different times, Chinese, Japanese and other non-white agricultural laborers (Chan 1987). The meaning of the land was decentered from the physical place even as it took on a new meaning as a symbol of the paradise and possibilities for the wealthy. In addition to serving as a fertile site for producing agricultural crops, irrigated with its vast supplies of surface water, the Delta soon became the center of the state's Romanesque plumbing system, the vast network of gargantuan pumps, aqueducts and canals that convey water over hundreds of miles from the northern areas to the dry and booming Central Valley and southern California. Such massive infrastructure to facilitate water transfer is a crucial element to California's "hydraulic society" (Worster 1985) and the cultivation of bureaucratic, hierarchical, and inequitable structures that such societies tend to produce. The development of water infrastructure normalizes the accessibility of water, first for urban sanitation, then for urban expansion, dual processes that some argue are central to the redistributive function of modern states (Kaika 2005:142).

Since California was ceded by Mexico under the Treaty of Guadalupe de Hidalgo and became a State in 1850, the Delta has been subjected to ever more aggressive attempts to control the hydrological and ecological processes that have shaped it for millennia, specifically the regular flooding from rain and the snowmelt from the Sierra Nevada mountains streaming in cyclic inundations onto the plain below, coursing toward the salt water estuary of the San Francisco Bay. The Delta is at the epicenter of California's large-scale engineering projects and technologies to "control nature" in McPhee's (1990) famous words. Beginning in the mid-nineteenth century, the "unruly waters" (Jones and Macdonald 2007) of the Delta had to be disciplined and made economically productive and rational. The Delta is a notoriously flood-prone system, with floods heavily damaging or destroying cities around the region (Isenberg 2006; Thompson 1957) and underscoring the unreliability of the water supply and the need for flood control (Mitchell 1996). While to early boosters the Delta represented nature's dangers to be tamed, eventually its fertile soils and easy access to abundant water, capital and markets made it an almost ideal site for agriculture. Indeed, between 1850 and 1868, federal and state legislation encouraged "reclamation" of swamplands and concentrated land ownership and wealth. The control of nature in the Delta was made possible through the labor exploitation of subjugated peoples, such as the Chinese whose labor "reclaimed" 88,000 acres of the Delta swampland for more "productive" uses by building levees and islands, dredging canals, and constructing small water projects to supply water for goods movement and drinking water for San Francisco (Arreola 1975; Chan 1987). The Delta began to resemble less a marshy, swampy wildland and more a collection of very large and productive agricultural enterprises bound by levees holding back the deluge (Wolff 2003).

For the last 80 years, the state's primary water policy objective has been to maintain the Delta as a freshwater system through water flow regulation supported by agricultural levees (Lund et al 2007). The growing need for freshwater by surrounding irrigated agriculture in the 1930s (Mitchell 1996) synched with the economic disaster of the

Great Depression. The result of this conjoining was the State Water Project, proposed in 1931, authorized by the Central Valley Act in 1933, and implemented as the Central Valley Project (CVP) by the federal government in 1935 (State Water Project 2008). What began as a flood control project was later used to provide water for agricultural and urban interests. The purpose of the State Water Project (SWP) approved in 1960 (and enacted as the Burns-Porter Act) was to provide flood control and water to urban interests, and to create a supplemental supply of fresh water that would control saltwater intrusions and compensate for diminished flows during peak water usage and north–south water transfers to southern California (Lund et al 2007; Mitchell 1996).[8] The Central Valley Project is one part of the SWP. Funded as a Depression Era project, construction of the California Aqueduct started in 1937 and continued until 1990. The CVP provides water from Shasta and Trinity Rivers to the Central Valley agricultural fields and urban centers.

The Delta was thereby transformed in the state's spatial imaginary from an unruly and flood-prone wetland to a vast water supply transfer node and conveyance system through giant feats of engineering (Mitchell 1996). Through its transformation, the Delta became a freshwater environment, a change that has had profound effects on the region's ecosystems. The water resources of the Delta thus became valued by the state of California as the Delta became the primary water conveyance structure to enable the large-scale urbanization of southern California and to feed capital, both industrial growth in the Bay Area and the agricultural sector in the San Joaquin Valley. In other words, the technological, environmental and political history of the Delta illustrates how more than a century of water policy changed conceptualizations of geographic scales, transforming the Delta from a local ecosystem to one that could be used and exploited in larger regional and state-wide extents. Thus, through public policy, modern engineering, and discourses of economic necessity, desires both proximate and distant have reached into the Delta, gripped its waterways and soggy soils, and bent them to their liking, diverting rivers, reshaping landscapes, and altering everything from microhabitats to the very current of the water via the power of the pumping stations.

For a full century beyond California's statehood, the twin forces of the state and capital interests dominated how the Delta was perceived and managed, organized around a view of nature to be dominated and controlled and its water resources exploited. This view was first challenged in the 1960s by the ascendant environmental movement and by a whole host of federal and state legislation to protect water quality, unimpeded flow, and endangered species (Little Hoover Commission 2005; Lund et al 2007). These new laws and public dissension over unchecked water use and disposal of natural resources halted further work on the SWP. The State Water Resources Control Board made various efforts throughout the 1970s and 1980s to require that some Delta water be used for environmental (defined as water quality) purposes; each attempt touched off a backlash of lawsuits by agricultural interests, and contention over rights to Delta flow-through intensified (Lund et al 2007). Perhaps the most incendiary of conflicts during this period was over the proposal for a "peripheral canal" as an alternative to the current water conveyance/pump system. The 1000-foot wide canal was designed to bypass the Delta, taking water from rivers in Northern California and diverting it around Sacramento directly to the San Joaquin Valley and points south. The peripheral canal was not built because a majority of California voters rejected the approval of bonds to build it in 1982. A peripheral canal has long been considered a third rail of California politics (akin to taxes), and this proposal has been recently resurrected by some interests as the only hope for balancing environmental and economic interests in the state (Lund et al 2007).

Controlling the Delta in the service of the economic growth of California depended upon particular goals of controlling nature and its risk (such as flooding) through technology and engineering. At the same time, human systems of injustice are central to this history, specifically the displacement and destruction of Native American populations and the labor exploitation of racialized populations (in both the engineering projects of Delta reclamation and the agricultural industry that emerged from its wake). These historical processes of social injustice, shifting geographic conceptions of Delta water, and environmental exploitation converged to set up the ecological conditions for its collapse. The next section examines how current policy attempts to fix the Delta—the Delta Vision process—recreate and exacerbate the very problems they aim to correct.

The Delta Vision and CALFED: Contesting the Politics of Scale, Process and Politics: 1980–2008

At a recent meeting of the Blue Ribbon Task Force (BRTF), the Delta as Place Working Group submitted a report with initial questions and recommendations about the future shape of the Delta as it concerned the residents and infrastructure in the region (28 February 2008). The Working Group proposed that the Delta be considered a mosaic of possibilities and disparate concerns rather than a uniform problem to be solved. The Working Group proposal was met with a stern admonishment from the Delta Vision leadership. The chair of the BRTF reminded the group to keep their thinking about the possible future of the Delta "rational and pragmatic", saying that "where you wind up is largely determined by where you start" (Delta Vision Blue Ribbon Task Force Meeting, 28 February 2008). Let us begin at this conflictive moment between the rational (and thus the implied "irrational" view) and this articulated notion that "where you wind up is largely determined by where you start". Where do we begin to understand how to rationalize the Delta, given the complex social, ecological, technological and political histories outlined in the preceding section? What is the Delta Vision? Where did the process emerge from? Who is involved? What politics are embodied in the Delta Vision process? Whose vision is implied in the naming of the Delta Vision itself? Do the environmental, technological, social and racial histories outlined in the preceding section make it into the frame and discourse of visibility in the Delta Vision?

Delta Vision emerges from a process that extended from the late 1980s through the 1990s to "fix" the problems of the Delta. This process is known collectively as the California Bay Delta Agreement (CALFED) and is comprised of a group of 18 federal and state agencies formed in 1994 to work on long-term solutions to Delta water problems (Jacobs, Luoma and Taylor 2003). CALFED's goals rested upon the idea that all interests in and needs for the Delta and its resources could be met without any of the main interest groups—water exporters, in-Delta users, environmental groups, agriculture—having to give up what they desired. The idea was that, "everyone would get better together" (Lund et al 2007:ix) through managing the Delta as a single unit. However, in trying to make everyone happy, no one truly was.[9] Conflicts over multiple water uses in the Delta were aggravated by a drought that lasted from 1987 to 1992 and resulted in cuts to water for all Delta users to preserve water flows; even these cuts couldn't stave off the listing of two Delta fish species under both federal and state endangered species laws (Lund et al 2007:ix). In 1992, Congress passed the Central Valley Project Improvement Act, which reallocated some water from the CVP to protect fish and restore ecosystems and also authorized water marketing and the sale of water among water users. This authorization of water marketing is not exceptional, but rather exemplifies the ascendant moment of neoliberalism of water management and capital in the context of privatization and deregulation in the 1990s. As Kaika (2005:154) describes in her account of water politics in London and Athens, "casting nature as a source of crisis and defining water resources as scarce provided the context in which the dominant neoliberal rhetoric and attitude towards water resource management (ie demand management through pricing) could be further applied".

In the midst of the legislative and legal chaos around the Delta, the three-way water rivalry among agricultural, urban, and environmental interests began to subside when the groups came together in a collaborative truce. During this period, multiple federal agencies began an alliance to coordinate their management of the Delta—it was dubbed "ClubFed" by its members. Commodification of one form of nature (water supply) at the expense of another (fish populations and ecosystem function) was theoretically balanced by mitigation activities intended to protect the natural system. For example, fish screens and mechanical movement of whole populations of juvenile fish was carried out in order to reduce harm to endangered species caused by the massive hydraulic pumps. Furthermore, those deemed "third parties" without a direct economic or political standing (eg the low-income and communities of color effected by the management decisions) had little to no representation in the process (for a view of environmental justice under CALFED, see Shilling, London and Liévanos forthcoming A). Since the signing of the CALFED Record of Decision in 1996, the Delta has fallen into further disrepair, its fisheries are failing, its native species being pushed out by invasive species, its levees crumbling, and its water quality diminishing. Although there is no causal connection between these collapses and the existence of CALFED, neither is there evidence of improvement. Disagreement among interest groups over causes of these failures came to a head in 2004 and 2005 with legislative budget cuts, gubernatorial audits of CALFED's governing and financing structures, lawsuits by environmental groups, and Hurricane Katrina, which called into serious question the safety of Delta levees (Little Hoover Commission 2005; Lund et al 2007). Finally, the withdrawal of Federal agency funding and participation has pulled apart the Cal and Fed components leaving the state agencies alone to collaborate among themselves and with environmental interests.

CALFED made little headway and is being bypassed by combatants in the water wars, primarily Delta Vision. During the 2005–2006 California legislative session, two bills (along with an executive order) were passed and signed by Governor Arnold Schwarzenegger to deal with the failure of state and federal governments to improve Delta conditions. These actions required a strategic Delta Vision and the creation of a comprehensive Delta plan, facilitated by the BRTF.[10] According to their website, the mission of the Delta Vision is to "identify a strategy for managing the Sacramento–San Joaquin Delta as a sustainable ecosystem that would continue to support environmental and economic functions that are critical to the people of California" (Delta Vision 2008). The BRTF is advised by four working groups, which focus on how to plan in the context of the "co-equal" goals of the Delta Vision. They are: Delta as Place, Water, Governance/Financing and Estuarine Ecosystem. This Task Force created a "durable vision for the sustainable management of the Delta" that was delivered by the Task Force to the Governor in January 2008, with a strategic plan that is to be delivered by November 2008 (BRTF 2008).

To examine the environmental justice implications of this "vision" we will begin with a seemingly simply, yet incredibly vexed question: what is the scale of the Delta embedded within Delta Vision, and how does the Delta Vision process and its preferred choice of regulatory and geography scale exacerbate environmental and social injustice? To clarify the multiple scales at play, we draw upon Kurtz's (2003) description of three scale frames adopted by environmental justice activists in a particular struggle against a proposal to site a controversial polyvinyl chloride production facility in Louisiana. In her formulation, the local environmental justice movement use what she calls "scale-oriented collective action frames". She defines three discursive and political constructions of geographic scale: scale of regulation; scale that legitimates inclusion and exclusion in political debates; and scale as analytical category of academic and bureaucratic projects. Thus, following Kurtz, we trace how these scale frames are mobilized in our case.

Scale Frame 1: Scales of Regulation: Defining and Regulating the Delta

Scales of regulation are domains for spatial practices such as the setting and operation of jurisdictional boundaries, through which the state incorporates and regulates the Delta. The Delta Vision comes out of a history of state attempts to regulate and manage the Delta for interests at the state level, a particular geographic, regulatory and political position. In 1959, the California Legislature passed the Delta Protection Act, setting the legal boundaries of the Delta (Figure 1) and requiring future engineering efforts and water appropriations to consider the water quality of the Delta in order to maintain the viability of agriculture within it (Lund et al 2007). This spatial bounding of the Delta in 1959 in legal terms has significant regulatory and political impacts.

The Delta Vision process builds in scale conflict most clearly between the question of whether the Delta is a site through which natural resources flow, in this case water, or whether it is an integral and valued ecological and social unit *itself*. Though the scale at which problems are defined and solutions for them are sought has been ostensibly the legally defined Delta, in reality and for decades, the problems and potential regulatory solutions extend well outside this spatial extent. In addition to the Legal Delta, the Delta can also be understood through wider geographic contexts, such as ecosystems that interact with the Delta including the Sierra Nevada, Northern Coast Ranges, and areas that receive Delta Water in the Delta "solution area" (Figure 2). Although state regulatory processes such as Delta Vision and its predecessor, CALFED, have acknowledged the presence

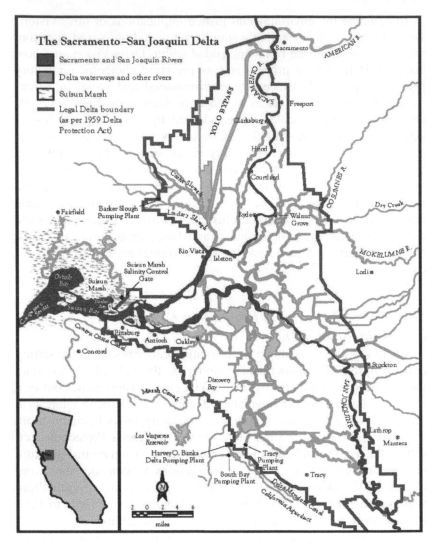

Figure 1: Sacramento–San Joaquin Delta (inset: Delta region within the State of California, from Lund et al 2007) (source: map by Janice Fong, UC Davis Geology Department)

of these broader geographic areas, these processes have focused on the Delta itself and its immediate tributaries without supporting or coordinating with activities in the broader areas (BDPAC 2007; Little Hoover Commission 2005). By focusing attention on the Delta and its immediate tributaries for mitigation solutions, the state obscures and avoids conflicts at the broader scale of the water source and use areas. This choice of the more legal Delta as the scale of regulation ensures that certain mitigation decisions render the sources of pollution and scales

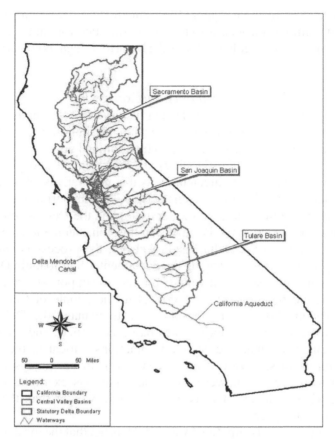

Figure 2: Watersheds and major waterways leading to the Delta (source: State Water Resources Control Board)

of water consumption, extraction and usage irrelevant and illegible to the process. In reality, however, they are central to understanding and solving of the problems within the Delta ecosystem.

For the Delta Vision process, as with CALFED, the choice of the Delta as a regional unit of analysis is the political and geographical construct through which decision-making is intended to "flow". The legal problems were identified at the scale of the Delta (e.g. declining fish populations) and water being conveyed through the Delta. A technocratic and scientific core of institutions developed around this ideological construct, including regional scientific conferences, a scientific journal, and multi-million dollar studies and ecological restoration projects (including land acquisitions and restoration along the Sacramento River riparian corridor, on specific Delta islands, and water acquisitions to benefit fish; see BDPAC 2007). Some of the more complicated environmental problems requiring investigation (eg agricultural intake and discharge effects on aquatic life) remained the least analyzed and received minimal financial resources from the state for study, arguably because the scale of regulation limits the Delta to its legal boundaries.[11]

Scale Frame 2: Scales of Inclusion/Exclusion

One of the central features of the Delta's history as a socio-natural space is its embodiment of the conflict between the state-building apparatus for capital development, and its status as a functioning and vital ecosystem, whose natural cycles (like flooding) are inconvenient and physically dangerous to cities and threatening to the imperatives of state and capital. This contradiction between the Delta's social purpose (to supply industrial agriculture and urban water users) and the Delta's ecosystem functions is reflected in the Delta Vision mission statement. According to the overview on the Delta Vision website, the process is aimed to "identify a strategy for managing the Sacramento–San Joaquin Delta as a sustainable ecosystem that would continue to support environmental and economic functions that are critical to the people of California" (Delta Vision 2008). Thus, the goals are to manage the Delta as a sustainable ecosystem (a local, place-based construct), and second, to support environmental and economic functions critical to the people of California (a statewide political construct). In our analysis of the Delta Vision process, however, we find numerous examples of how the statewide political interests hold far more weight than the local, place-based analyses.

For instance, the assumption by the BRTF is not that the local does not matter but that it only matters as a place for mitigating impacts of water management—it is not seen as a place for decision-making. According to the public discussions of the BRTF, the Delta and its massive ecological and infrastructural issues must be dealt with by agents acting on a regional and state level and regulatory scales, especially in the face of climate change, if the "co-equal goals" of ecosystem protection and water supply are to be met. In a recent public meeting the chairman of the Task Force said, "the co-equal goals will become measures of future efforts, even if it's not what people in the Delta would wish". In other words, the local view is not important to consider in decision-making. To further emphasize this view of the local, the Chair said, "the Delta as Place Working Group must look at the whole vision from the state perspective, which is not a comfortable way for Delta interests to start" (Delta Vision BRTF 28 February 2008). The Task Force chairman views the Delta Vision process from the scale of regulation of the state of California. This view of the state as the appropriate and preferred scale of regulation creates conditions of relative inclusion and exclusion in the political process, with interests deemed as "special" (ie local) being marginalized relative to more abstract interests of the state and the ecosystem itself. This view of the state scale of regulation is the dominant perspective adopted by the Task Force in creating a strategic plan that foregrounds maintaining water for economic uses for agriculture and for water transfer. In other words, the BRTF suggests that the Delta should be a placeless space that fulfills the twin goals of ecosystem health within the Delta and water "for the people" outside the Delta proper. The dual narratives of state building and "the-good-of-the-people" (abstracted as non-local citizens of the state) is repeatedly invoked throughout the Delta Vision Process. These narratives are reinforced by major regional scientific bodies, exemplified by a recent report (Lund et al 2007), which describes the parameters to consider when choosing management and restoration options for a future Delta. These parameters include: "flooding, predicted sea level rise; water quality and the recognition that the new ecosystem will be a different ecosystem than the one we currently have. This means that there is a unique opportunity to rebuild the ecosystem into one with attributes that society decides are desirable. By recognizing that the ecosystem will undergo major change with or without human intervention, it is possible to capitalize" on the opportunity of the Delta decision-making process in response to the crisis in the Delta (Lund et al 2007:56).

The Delta as Place Working Group is, by accident more than design, the work group where all issues that do not fit into the other more politically powerful work groups (Water, Governance/Financing and Estuarine Ecosystem) are allocated, however awkwardly. Thus, the wide-ranging topics addressed by the Working Group include: land use and infrastructure; levees and flood plains; emergency management and response; recreation and tourism; and transportation and utilities. Despite this agglomeration of issues, the Delta as Place Working Group has attempted to conceptualize the variations in needs, cultures, and landscapes within the Delta in order to come up with a plan that will forward the desires of in-Delta interests (Delta Vision BRTF 28 February 2008). The Working Group allows for discussion of the Delta as a spatial patchwork of possibilities in flux. This allowance of flux acknowledges ecological and cultural variability and contrasts squarely with the politics and ideology of the BRTF writ large. The BRTF embodies a state-scale of regulation approach that frames a more abstracted problem needing a more across the board and aggressive response, especially in advocating engineering solutions, and through a relatively closed political process.[12] The Delta as Place Working Group also imagines the Delta in a different temporal scale than the BRTF. The Working Group discusses and makes legible the immediate and longer-term impacts on the actual people who live in the Delta. They have invoked the narrative of the "working landscape" as a way to construct their own Vision for the Delta, as well as proposing several regional governance structures to deal with the varied needs of the southern, central, and northern Delta's islands and unique ecologies and communities. In contrast, the BRTF, and its state-scale of regulation approach, desires predictability, consistency and control, a view of nature, places and people at odds with the Delta as a dynamic ecosystem or a place where humans matter and where social and environmental injustice take place.

This discursive conflict between the local and the state is how the construction of geographic scale maps the scale of inclusion/exclusion onto the political process. In other words, local interests are constructed as parochial, short sighted, and irrational, whereas advocates for the state are considered to be acting for the abstract good, rather than in the service of capital and large water users (agriculture and urban water districts). What then becomes of Kurtz's scale frame of scales of inclusion/exclusion, and the question of environmental justice itself? While the Delta as Place Working Group does not explicitly invoke environmental justice in its mandate or its vision, its focus on broadening the bounds of the Delta geographically and epistemologically, as well as their sensitivity to spatial/temporal variability and the dynamics of inclusion and exclusion, offers an implicit conceptualization of environmental justice in the Delta. It is to this contested representation of environmental justice in the Delta that we now turn.

Scale Frame 3: Scale as an Analytic Category: Defining Environmental Justice in the Delta

Recent literature has disputed whether environmental justice movements and the academic evaluations in the literature are politically reformist rather than radically transformative. As Schlosberg (2004:518) points out, distributive justice ignores "the social, cultural and symbolic, and institutional conditions underlying poor distributions in the first place". Distributive justice, a central element of environmental justice claims, limits the possibility of radical confrontation in that environmental justice discourse can serve to promote the distribution of resources in a way that creates a veil of appeasement among those with limited access to the distributed resources. The claim for fair distribution suggests that if everyone receives equal access to the distribution of resources in a society, justice is achieved, thus eliminating conflict over resources themselves. In addition to the focus on distributive justice, the environmental justice movement has emphasized a procedural sense of justice, claiming representational space in the political and policymaking arena and the right to "speak for ourselves" (Cole and Foster 2001). To these questions of distributive and procedural justice,

Schlosberg adds that "recognition" of diverse cultural identities in a critical pluralism is a pre-condition for entry into the distributional system and ought to be considered a third definition of justice in environmental justice.

Questions about the nature of environmental justice claims include theoretical questions taken up by critics of environmental justice movements (Harvey 1996; Swyngedouw and Heynen 2003) who consider environmental justice to represent reformist and primarily local claims. However, environmental justice analyses of the Delta are not necessarily politically reformist, narrow, or geographically local. This critique of environmental justice is echoed by the Delta Vision BRTF, which framed the local—and, by extension, environmental justice—as a narrow and parochial "special interest". During the final stakeholder meeting for the Delta Visioning process, for example, when questioned about the lack of mention of environmental justice principles or goals in the final Delta Vision product, the executive director of the BRTF said that it was the decision of the Task Force not to include "technical or special language" from any stakeholder group in the final Vision (Delta Vision Stakeholder Coordination Group Meeting 17 December 2007). This interpretation of community needs and environmental justice issues as "special interests" particular only to *them* downgrades the scale of their argument from the accepted state/Delta scale to the merely local. Environmental justice has been relegated to a small and—in the eyes of the decision-makers—a "special case", therefore, non-includable entity in the Delta Vision (Water Education Foundation Delta Vision Workshop 7 March 2008).

What exactly are these environmental justice issues and communities in the Delta? Like the seemingly simple, but exceedingly complex nature of the Delta, environmental justice is a highly contested term. There are at least three interconnected levels at play, although the last is the least understood: the policy/legal, the activist, and the theoretical, in the view of socio-nature (which we analyze in the next section on the socionatural Delta and environmental justice).

First, California has a number of state laws that explicitly define environmental justice (London, Sze and Liévanos 2008). By these policy measures, the Delta Vision's characterization of the requirements for the Vision likely violate state law requiring that state agencies act to prevent disparities in harm and to facilitate participation of affected parties in decision-making. This interpretation of what is required of the state ignores the Record of Decision (ROD) that theoretically provides the state with the legal right to cause harm to the Delta. In the ROD, the state must include environmental justice concerns and communities as defined in state law in implementation of all CALFED program plans.

Second, activist conceptions of environmental justice in the Delta have been highlighted by advocacy organizations such as the Environmental Justice Coalition for Water along with other environmental justice groups. These problems include higher than average rates of mercury contamination in Southeast Asian and other racial minority communities (related to consumption of local fishes), and access to poor-quality water in farmworker communities as the result of pesticide runoff, and limited in-community water infrastructure. Large percentages of Southeast Asian refugee populations living in and around the Delta face elevated mercury contamination risk (a legacy of the state's gold mining practices), as a result of subsistence fishing practices that lead to their high rates of fish consumption from the polluted rivers (Shilling, White and Lippert forthcoming; Silver et al 2007). Advocates such as the Community Water Center (2008) highlight the cruel irony of poor water quality and lack of water infrastructure in poor, primarily Latino rural communities. In other words, clean water comes through the California Aqueduct to feed the thirsty fields of industrial agriculture, bypassing farmworker communities who drink from water contaminated by nitrates and pesticides [Environmental Justice Coalition for Water (EJCW) 2005]. As one environmental justice advocate identified:

> you've got these canals with pristine Delta water flowing through them, and literally they're 100 feet from a drinking water well that's supplying communities with nitrate-laden drinking water. So there, these communities don't have safe drinking water while this great water is literally flowing along through their backyards (interview 28 July 2008).

In response to these problems, a number of organizations have engaged in a range of activities, from protests, to educational forums. On the issue of mercury contamination, a number of events were organized by groups. For example, in 2008, community meetings composed of the Lao, Mien, Hmong and Russian communities were held. These meetings were translated into Lao, Hmong, Mien, and Russian and consisted of exchanges of technical information and concerns about environmental contamination among academic, community organizations, and community members (see Figure 3).

The EJCW is the organization that has most consistently tracked environmental justice with respect to water policy in California. The EJCW is a network of more than 50 grassroots and intermediary organizations that helps to "empower people to advocate for water justice in their own communities and assures policy makers listen to the concerns of those local community members" (2008). The EJCW membership ranges across California and is not solely devoted

Figure 3: Sacramento community meeting, 2008 (source: photo by Fraser Shilling)

to matters within the Delta. It is not a local in-Delta organization, and it is, therefore, an example of the "scaling up" of environmental justice advocacy to a regional and state-wide basis. The EJCW defines "environmental justice communities", as exhibiting at least two of the following three criteria: economically disadvantaged; disproportionately composed of people of color (defined as higher proportion of color than state percentages); and disproportionately impacted by environmental hazards.[13] The EJCW also articulates their work through a frame of "water justice", which they define as "the ability of all communities to access safe, affordable water for drinking, fishing, recreational and cultural uses".[14]

The EJCW has been publicly critical of the Delta Vision process, and what it calls in a letter to the task Force, its "ill-conceived notion of environmental justice in Delta Vision deliberations", which essentially ignores the existing state legislation governing environmental justice, specifically the legal language and moral responsibility to protect socially vulnerable and environmentally overburdened communities.[15] At the same time, the EJCW has continued to be engaged with the process, primarily around educating others about how environmental justice is not just a narrow set of issues (or "special interest"), but a cross-cutting analytic framework that should be incorporated across the Delta Vision process. The 2 September 2008 EJCW letter to Delta Vision makes a number of recommendations for an environmentally just Delta Vision, including: disciplinary and credential diversification within the Delta science program; increasing disaster preparedness and enhanced flood protection for socially vulnerable and environmentally overburdened communities; democratization of Californian water for socially vulnerable and environmentally overburdened communities; ensuring that water supply quality and reliability is carried out to provide safe, affordable water for drinking, subsistence, cultural, and recreational uses for California communities; promoting just and sustainable local and state economies and land uses within the Delta that are connected to vital drinking water sources throughout the state.

In sharp contrast to the Delta Vision process and its rejection of local and/or so-called special interests, the EJCW conducted its own research project on environmental justice in the Delta, which culminated in a 2009 report entitled "Third parties no more: Envisioning an environmentally just and sustainable Sacramento–San Joaquin Delta" (Liévanos et al 2009).[16] The report focused on the socially vulnerable and environmentally overburdened communities that the Delta Vision ignores, specifically highlighting the plight of racial minorities and economically disfranchised communities. It addresses the potentially disproportionate impact of flooding and other hazards and their differential impact on socially vulnerable populations (Cutter 1996; Cutter, Boruff and Shirley 2003).[17] The report exposes the power dynamics of the Delta Vision process itself and how it disregards the concerns of low-income communities and communities of color, including fears about loss of land and community. It also reports how these same communities have had to pay inequitable fees for levee restoration and flood protection, to drink potentially compromised drinking sources, and to fish for subsistence in contaminated waters (pollution legacies from point and nonpoint polluting sources). For some of the indigenous populations of the Delta, the pollution, water diversions, and land development in the region have reportedly exterminated or seriously degraded the materials they use for traditional medicines and basket weaving. In terms of procedural justice, these communities have had little to no public input into the decisions giving rise to these conditions due to the history of backdoor deals in water governance and current biases towards sporadic town hall meetings and often English-only, internet-based modes of information dissemination.[18]

The specific realities of Native populations are a particularly poignant example of how the politics of distributive and procedural justice and cultural recognition are intertwined, yet deliberately excluded from the state scale of regulation through the Delta Vision process. As one representative from an environmental justice organization explained, because of the history of the California Water Project:

> dams and diversions have depleted the salmon populations that are the lifeblood of tribal, headwater communities. Those people really suffer. Some of the coastal streams are talked a lot about, but in the Bay-Delta system, they are not talked about. Those people feel they won't heal until the river is healed" (interview 2 July 2008).

Another interviewee from an environmental organization concurred that:

> north of the Delta, there are tribal impacts, as well as other impacts, when the Trinity River is diverted into the Delta. That has huge impacts on the Hupa tribe, and the Shasta Reservoir has significant impacts on the Winnemem Wintu that don't even have recognition anymore most likely because of their proximity to the Shasta Reservoir (interview 8 July 2008).

The Winnemem Wintu have lived in a valley where four rivers meet, and have fished and farmed for centuries at the confluence of these rivers. Much of the tribe's ancestral land in Northern California was submerged when the federal government built a 602-ft dam downstream of their ceremonial and prayer grounds in 1945, with the rest facing inundation if a proposed US Bureau of Reclamation to enlarge Shasta Dam as a way to boost California's water supply is approved (Egelko 2008). Indeed, the comprehensive study plan covering raising Shasta Dam and inundating the historical and sacred lands of the Winnemem Wintu provides just two sentences to describe the ability of the tribe to provide input, but no official recognition that the lands or tribe will be impacted (US Bureau of Reclamation 2007).

In addition to Native concerns (which are arguably distinct from environmental justice frameworks because of the unique history and land sovereignty issues), questions about the existence of environmental justice communities, and exactly where they are located in the Delta are complex affairs. In both the Delta Vision process and our supplemental interviews, particular towns or neighborhoods did not come up as having specific environmental justice issues. Rather than being place-based or local communities, "environmental justice" instead seemed to be loosely defined by interest or issue groups who depend on the Delta for economic and/or cultural purposes. These everyday populations include migratory and settled agricultural laborers; people dealing with poor water quality or disparities in water deliveries; groups facing a loss of heritage due to regional policymaking; subsistence fisher groups whose food sources are diminishing due to ecosystem collapse or poisoning by mercury and other toxins; people living in urban-fringe and in-Delta areas that face the real possibility of inundation due to sea level rise, seasonal flooding, or levee failure, and who are at greater risk because of uncoordinated emergency planning efforts by counties and municipalities; and people who must deal with poverty and economic underdevelopment because of skewed land-use priorities (Fielding and Burningham 2005). In other words, environmental justice is not a specific group of people or discrete pollution sources, but a framework of analysis, grounded on articulating the connections between racial and social injustice with ecological problems.

During the Delta Vision Process in 2007–2008, environmental justice advocates had a small but earnest trio of representatives at the Stakeholder Coordination Group decision-making table. According to one of these individuals, the group did their best to educate the rest of the assembled stakeholders (including mainstream environmental groups, recreation groups, in-Delta town leaders, water exporters, agricultural groups, and Southern California water importers) about the immediacy and relevance of environmental justice issues to every other issue brought into the discussion. One of our informants felt that the trio of environmental justice advocates had begun to make significant inroads into alliance building with formerly hostile stakeholders, although their success in persuading the BRTF was minimal (interview 28 July 2008). During the Delta Visioning process, in which 43 officially recognized stakeholders were allowed to collectively create their own vision for the Delta, a document was produced to elucidate their vision. Some of it was eventually incorporated into the BRTF's Official Vision document (Delta Vision Stakeholder Coordination Group 2007), but the amount of attention their vision was given is subject to interpretation. In addition, individual organizations submitted their input, under the rubric of "External Visions".

In this context, the EJCW articulated their "Environmental Justice Vision" for the Delta (EJCW 2008). This Vision included key procedural justice elements, such as enhanced capacity for participation in the process, incorporation of "meaningful stakeholder engagement", identification and correction of data gaps relevant to communities, and a requirement that decisions made with inconclusive data be made reversible and provisional. Their Vision also included provisions for adequate drinking water quality and supply for all state residents and for removing methylated mercury, which poisons fish and adversely affects subsistence fishers. Environmental justice advocated consideration of the needs of low-income Delta residents: during processes of land-use change their property would not lose value or be irrevocably lost in emergency situations; environmental justice communities would not be disproportionately affected by increased flooding risk due to land-use changes; and emergency response mechanisms would be in place. It stated that disadvantaged communities should have access to economic development opportunities and that adverse economic impacts due to the loss of the Delta's agricultural base should be considered. The environmental justice vision also recognized the impacts of upstream flood and flow control on the health of the Delta. Finally, it also called for adequate flows of water and future flexible responses to maintain Delta ecosystem health, and the importance of processes besides flow on the health of the Delta. In short, an environmentally just Delta Vision, according to EJCW and other environmental justice activists, would include means to address distributive and procedural injustices as well as ways to incorporate cultural recognition into the planning and governance processes. Cultural recognition incorporates the worldviews and histories of disenfranchised and subaltern populations, and the history and cultural practices of Native populations in the Delta. The EJCW Vision used both "narrow" scales of community and individuals and "broad" scales of ecosystem, economy, and emergency management to construct a comprehensive, but specific list of the needs and concerns that the Delta Vision might address.

Environmental Justice and Socio-nature in the Delta

Environmental justice advocates within the Delta Vision process were marginalized. Despite this, their collective efforts represent some of the most vigorous efforts within the process to take human populations seriously and from a standpoint that is not in the service of capital and the state (agricultural and urban water interests). Environmental justice advocates attempted to reconstruct the framework in which the stakeholders were placed as something fluid, shared, and truly collaborative. This can be construed as an attempt by a marginal perspective to claim the center of discourse and policymaking. Nearly every group participating in the Delta Vision process referred to it as "Delta Vision", discursively denoting a single "vision" even if in their minds they had constructed their own narratives about the region's utility, meaning, history, and best future uses. In our fieldwork and interviews, only one respondent, a member of the environmental justice advocate group, repeatedly invoked the process as the "Delta Visions", thereby acknowledging the simultaneity and possible confluence of all the groups' desires. Environmental justice advocates best articulate the need for these "multiple visions" because of their discursive focus on interconnections, with a strong focus on socially vulnerable and racial minority populations (Cole and Foster 2001). That is, environmental justice analysis, as a political script, parallels the critique of socio-nature, which rejects the formulation of the society–nature or technology–nature dualism. But, because of the structure of the Delta Vision process, environmental justice is primarily articulated by activists pressing the state to fulfill its statutory obligations for addressing environmental justice, which are still reformist rather than radical. This tension between the reformist and the radical is yet another example of Audre Lorde's poignant question as to whether the master's tools (in this case, the Delta Vision process in particular, and regulatory action more generally) can dismantle the master's house or, in this case, water networks. For example, the EJCW focuses primarily on disparities in access and quality, not exclusion from and lack of ownership of the decision-making process about water. Even the EJCW view of the Delta does not incorporate a theoretical view of the Delta based on a socio-natural critique. That said, environmental justice analyses in the Delta, primarily (but not exclusively) represented through the EJCW report, suggest that discourses of justice can complicate and potentially be a policymaking process by highlighting the power of the state and of capital to exacerbate conditions of social injustice that have ecological effects in the Delta, primarily through telling different stories.

This articulation of the possibility of multiple visions challenges the political fictions that underwrite the script of the Delta Vision: that co-equal and separate domains of economic and ecosystem exist, and that that there is a single vision that would enable both to thrive. In reality, economic and ecosystem functions are thoroughly embedded within one another, in a complicated agglomeration of socio-natural and techno-natural spaces. This critique is especially needed in light of the re-emergence of large-scale engineering solutions, particularly the Peripheral Canal, as solutions for the problems that past engineering infrastructure helped to create.

Any future action in the Delta that ignores the environmental history of the region, and the critique that engineering solutions have created many of the problems in the Delta, will probably lead to a costly repetition of past failures. Engineering solutions depend upon economic and utilitarian analyses that are deeply ideological, even if understood by their practitioners and advocates to be apolitical. One academic scientist we interviewed understood environmental justice in these terms:

> If you do not choose the Peripheral Canal, pretty much everybody else loses...the biggest hit goes to the San Joaquin Valley [agriculture]... with the biggest impact onto essentially poor working, the impact on the poor, over 100,000 [jobs] (interview 5 August 2008).

In other words, in his estimation, unless water supply to industrial agriculture is stabilized, the loss to farm labor was the main environmental justice problem. Similarly, another university engineer conflated environmental justice with economic concerns, and dismissed health concerns about water quality as largely "unsubstantiated" (interview 21 August 2008).

In our analysis of the Delta Vision process, there are several dominant assumptions: history does not matter to public policy; economic impacts are distinct from the ecosystem (which humans do not inhabit); and any local scale or discursive frame of environmental justice is seen as "political" and suspect. In contrast, the state scale of regulation is seen as effectively possessing a sufficiently broad geographic and regulatory scope to "fix" the Delta. These assumptions also ignore the notion that, just as we create socio-nature in the form of the Delta, the Delta also influences. Here, the literature of socio- and techno-natures is particularly useful. This perspective highlights how people, places and things are technologically mediated, produced, enacted, and contested (White and Wilbert 2009). In sharp contrast to the Delta Vision process, we see the Delta as a hybrid waterscape of socio-natural production (Swyngedouw 1999). The lens of "socio-nature" allows us to illuminate the politics of the Delta's transformations. In apprehending the Delta as a hybrid socio-nature we have attempted to present a vision of the Delta that highlights the ways in which the control of water is bound up in the control of human populations and economic processes, and how such control is central to the modernization project of the state. Drawing on Robbins (2007) we suggest that the agency of non-human nature, expressed by its unruly and inconvenient floods, species movement, and feedback loops is an integral part of this socio-natural production. We defetishize the human labor that invested itself in this production. That means excavating the efforts of the Chinese laborers who built the Delta's levees, highlighting the lives of the agricultural laborers that worked the Delta's soils, and who now increasingly build sprawling suburbs where orchards and row crops once reigned supreme. It is in this everyday Delta, this historical Delta of lived experiences where the abstractions of the Delta as an ecosystem and Delta as an economic engine fall away and a more complex reality emerges. It is here that a vision of an environmentally just Delta can begin to be articulated.

Conclusion: Imagining Ecological and Environmental Justice in the Delta

Overall, our research shows that the master narrative currently coalescing around the Delta Vision suggests that there is no way to incorporate people into the state-sanctioned vision of what the Delta should be without compromising the goals of that vision—the co-equal goals of undertaking economic development and maintaining ecosystem functions. Local people within the system are a "distraction", to borrow the language of one of our respondents (interview 5 2008). They both (people in general and racial minorities in particular) make it harder to recommend and embark upon desired courses of action because they add messiness and contentiousness to the process, as does injecting considerations of justice. If they and their issues are recognized, they will demand to be included, to have their concerns met, to have representation, to receive equal consideration, and to work towards justice—all of which are protected by a number of state and federal law and executive orders (London, Sze and Liévanos 2008).

Our findings are that the populations and communities in the Delta, who themselves possess little political or economic power but who bear the brunt of a number of (state-sanctioned and state-caused) problems, are largely excluded from regional decision-making processes in the institutionalized policymaking arena (as in the Delta Vision) and in efforts to combine science—defined in particular ways and by particular people—with policymaking. The Delta Vision process has, by seeking to "balance" the co-equal goals of protecting the Delta as an ecosystem and the Delta as a water conveyance mechanism posited these elements as separate, and thus irreconcilable. Therefore, in the very formulation of the problem lies the obstacle to its resolution. Furthermore, by taking up the entire space of the policy debates, these two goals have restricted questions of social equity and environmental justice to a small part in the drama and incarcerated them in a parochial rendition of the local. Indeed, even the potentially libratory discourses of environmental justice are consigned to the narrow view. There is no space, therefore, for a radical critique of the ways in which the "nature" of the Delta has been transformed by social, economic, and political systems far outside its boundaries; how the scaling and representations of the Delta in public policies designed to "save" it reflect these dominant interests; and how the framing of the Delta within these policies erases or negates the voice and interests of subaltern populations inside and outside the region.

  These results are not unique to California's water planning processes; however, they are significant, in part, because of how they impinge upon the public image California has constructed for itself—a place of liberal and inclusive ideals, a state where all are welcome to create and indulge in the opportunities that abound within its borders. While the state weaves a story about the importance of the Delta as being "indispensable to a modern California" (BRTF 2008:3), it excludes many of those it purports to welcome from participating as modern citizens. This research offers a reminder of the mechanisms in the state's employ that allow it to protect its economic interests. These include using the politics and narratives of scale to make invisible the localities that are most useful to the state as resources for, or drivers of, industry and profit, and the populations living in those localities who might offer counter hegemonic claims and counter narratives that, if allowed to enter the public discourse, might displace the state's claim upon the resources it desires (Desbiens 2007). In many ways, the state has effectively separated itself from the interests of the people, as expressed by the people themselves, instead aligning itself with a conceptual framework of agricultural and urban water needs, as expressed by water agencies. The Delta as a highly constructed socio-nature has become integral to the collective vision that the state has created for itself and, by extension, for the entire population of California. Without maintaining control over the Delta and its future material and narrative constructions, the state faces losing its own identity as a modern force for progress, industry, agriculture, and urbanization. Therefore, it may be imperative for the state to construct the "population" served by the Delta Vision process (and therefore benefiting from an "improved" export system and ecosystem) as the collective body of all the state's residents, rather than as individual groups with conflicting needs and desires. Constructing them as irrational, illegal, or illegitimate, as the state has done to the north-of-Delta tribe, the Winnemem Wintu, allows the state to maintain control over the conversation about who is a legitimate stakeholder in this grand, modern re-engineering water project that is the Delta Vision.

The cultural and ecological discourses employed by the Delta as Place Working Group and the EJCW, or by Native American tribes who experience the Delta through cultural practice, counter the Task Force's choice of scale or view of nature. Each is seen by the state and capital, through the Delta Vision process, as obstructionist. These discourses are not seen as adding texture and nuance to a plan that will shape the future of all in-Delta residents; they bring the rest of the state's notice and concern to the Delta's resilience and longevity. The Delta Vision process very consciously builds in scale conflict, ignores different visions of justice, and depends upon views of nature that, as befitting its larger ideological underpinnings, are in the service of the state and capital. Intentionally inappropriate scales of scientific study and managed solutions have led to massive disconnects between the semblance of public process and the actuality of water management decisions.

Water management in California, the Central Valley, and the Delta has suffered from decades of control by forces opaque to the public in whose name many decisions are made. The structures of governance, alongside commodification of nature and extraction of water, have brought the ecosystem to the brink of collapse. Historical and contemporary water management policy in the Delta and throughout California ignores cultural experience, perspectives and insights of ecological sciences, and is profoundly ahistorical. Instead, an ecologically just Delta planning process might imagine and re-envision the region as actively connecting and re-connecting with people, place and water, at complex and diverse geographic and temporal scales. It would also recognize multiple forms of knowledge, acknowledge the socio-natural dimensions of the Delta rather than treat the economic and ecological dimensions as separate, view the Delta as a problem to be solved, and no longer dismiss alternative views of Delta planning as "special interest", while the broader scale of decision-making is represented as rational, rather than itself interested in protecting the forces of state and capital.

In other words, environmental justice and scale *do* matter in the Delta in rather complex and profound ways. Environmental justice in the Delta necessitates an understanding of the interconnections between distributive injustice, procedural factors, and the politics of cultural recognition, and an understanding of space and scale as geographic, sensual and temporal. This approach is a radically different approach than the status quo in which the state characterizes the needs and desires of socially vulnerable populations in California as local and parochial at the same time as advocating forcefully for capital and urban interests. In summary, an ecologically just process of decision-making for water in the Delta would: include in all scales of decision-making the human communities impacted by delivering water (from source areas), conveying water (through the Delta), and using water (in urban and agricultural areas); redistribute power over water management to non-traditional stakeholders, including those representing natural systems; distribute the entire costs of conserving water, conveying water, and mitigating the impacts of water management to those who have benefited fiscally; eliminate the decision-making bodies and processes that have to date led to the conflicts and impacts over water in California, as described here and in previous studies; and set as a goal, both an ecologically restored Delta and watershed, and restoration of connections between people, places, and water.

Acknowledgements

We thank the Committee on Research at UC Davis who funded the research under the auspices of the Collaborative, Interdisciplinary Research Program.

Endnotes

[1] From http://deltavision.ca.gov/AboutDeltaVision.shtml

[2] Acre-feet is a measure of water volume. It is the amount of water needed to cover an acre of land to the depth of one foot. 1 acre-feet = 325,851 US liquid gallons.

[3] In probing the connections between science and Delta policymaking, we chose to conduct interviews both with people involved directly in the Delta Vision process and with people involved with organizations associated tangentially with Delta-related environmental and science-making issues. Some people were chosen from a list of participants in the Delta Vision Stakeholder Coordination Group: these interviewees were chosen based on their affiliation with science-making or sympathetic (environmental justice and environmental) policymaking groups or non-profit organizations. Other people were chosen by the principal investigators on our research team based upon their own ideas of who might be informative to talk to about the intersections of science and policy in the Delta. Subsequent interviewees were chosen based on recommendations from the first set of interviewees, through a snowball sample technique. A total of 13 interviews have been done to date.

Because of the sensitive political nature of the topic, we promised anonymity to our interviewees. Thus, when we quote from interviews we conducted, we refer to them by their primarily organizational identity and the date that the interview was conducted, ie agency scientist, interview, date. Respondents have included employees of Delta-relevant state agencies, environmental NGO employees, and university professors. An interview protocol was emailed to each interviewee ahead of the scheduled interview. Questions focused on individual respondents' definitions of science; their perceptions of the quality, thoroughness, and use of science related to the Delta and in conjunction with Delta policymaking; the spatial and scalar extents of Delta science; and the extent to which interested or affected groups have been included in the science and policymaking processes. Despite this structured approach, the interviews did not stick with the interview protocol verbatim and instead proceeded in a more informal, semi-structured fashion, touching on major points and leaving aside points about which the interviewee had little knowledge. Each interview was preceded by an explanation of our overarching research questions and purpose, and the reasons the current line of questioning was being explored. Interview times ranged between 40 and 80 minutes. All interviews were recorded on a digital voice recorder; digital files were transferred to a shared website and to individual computers and were transcribed. Transcribed interviews were coded using Atlas TI, a computer program that facilitates sorting and comparing quotes and codes among many transcripts.

[4] These meetings allowed us to become participant observers in the policymaking process, to talk with stakeholders and interested audience members—often state agency staff members, but also members of the public at large and of various stakeholder groups who also were represented "at the table"—and to question Task Force representatives publicly about the absence of environmental justice concerns from their policy discourse. On several occasions, when in-person meeting attendance was not possible, we watched webcasts of Blue Ribbon Task Force meetings, which were held monthly. While this online attendance restricted our ability to interact with and observe the audience and off-camera discussions among Task Force members, it still allowed us the opportunity to get an impression of the Task Force and its members' personal exchanges with one another and with staff and other presenters of information and updates on camera. During these online observations, notes were taken on the agenda and the control of the meeting by Task Force members; briefings by Task Force members; discourse between the Task Force and any outside presenters of information, such as state agency staff members; and discourse among Task Force members. Direct quotes of interest on the subjects of public participation, science and policymaking, Delta as Place, justice, ecology/environment, and non sequiturs were typed into notes, as was the context for their utterance.

[5] We use socio-natural and techno-natural interchangeably. White and Wilbert identify the "the technonatural" as a "nature regime" interacting with organic and capitalist natures in complex and contingent ways. "Techno-natures", is not simply referring to a material referent of emerging artificial natures but is understood as much as a cultural sensibility, a phenomenon of everyday life, an imaginative horizon and an ideology (White and Wilbert 2009).

[6] A number of agency and university researchers suggest that the Delta's fragile levee system inevitably will fail, that an earthquake, sea-level rise, or flooding due to future weather-based vagaries brought on by climate change will take many of them out, flood many of the farmed islands of the Delta, drown unprepared populations and recreational and agricultural sources of income, and taint the drinking and irrigation water for two-thirds of the state's population (Lund et al 2007).

[7] In particular, work from scholars from Science and Technology Studies and state formation is important to our analysis. For example, historical sociologist Patrick Carroll is currently engaging in work on the Delta. Previously, he has articulated the notion of a "science/state plexus" to describe the ontologically dense, interwoven, multifaceted, heterogeneous and yet intercommunicating nature of the connections between modern science and modern government (Carroll 2006).

[8] The two biggest features of the State Water Plan are the California Aqueduct (begun in 1960), a 450-mile long, 40-foot wide, 30-foot deep trench that conveys Delta water through the Central Valley over mountains to Southern California (Little Hoover Commission 2005), and the Oroville Dam and Reservoir on the Feather River which, in a "typical" year, stores and delivers 3 million acre-feet of Delta water for San Joaquin

Valley agriculture, and Bay Area and Southern California urban areas (Little Hoover Commission 2005).

[9] While the experiment was considered to be a "model for cooperation" by some (Koehler 1995), other analysts suggest that this collaborative model was in fact defined by uneasiness and conflict (Innes et al 2007), and chronic lack of resources (Raley 2005). Heikkila and Gerlak (2005:607) examine CALFED as one of their four case studies of large-scale collaborative environmental resource management, ultimately concluding that the data are not clear that collaboration is necessarily beneficial.

[10] The BRTF is composed of eight members, appointed by the governor; a steering committee of four state department secretaries and the chair of the public utilities commission; and a stakeholder coordination group composed of 44 members representing a variety of interests (environmental and environmental justice groups, tribes, water districts, sport fishermen, cities and counties, builders, growers, and others), with both interest delineation and representative members selected by the BRTF.

[11] In Koehler's (1996:51) otherwise generally supportive article about the CALFED process, she ends by noting that agricultural and urban water user interests were vastly more powerful than environmental interests in setting the research agenda and in receiving financial resources in terms of staff and funding.

[12] This preference for engineering solutions is particularly clear in the PPIC report, which advocates massive re-engineering of water management through and from the Delta, primarily through a canal that encircles the periphery of the Delta, which would have impacts to communities within and outside of the Delta (Lund et al 2007). These impacts would be felt through state budget burdens to build the infrastructure, lack of inclusion in the decision-making process, and displacement of local economies and communities to suit state-scale water management decisions. This report confirms an approach to managing water as a commodity through engineered devices, with limited control and mitigation of the impacts on ecosystems, and little input or oversight from communities at any scale. Although the BRTF did not endorse the PPIC report, it does call for something like the peripheral canal (but that isn't called the Peripheral Canal because it won't be exactly like the one proposed in the 1980s), the BRTF is recommending the non-peripheral canal/peripheral canal as something to pursue (see volume 2, showing revisions, pp 45–46 of the pdf; http://www.deltavision.ca.gov/ StrategicPlanningDocumentsandComments.shtml)

[13] EJCW usually references the California EJ statute (SB 115; introduced by Senator Hilda Solis 1999): "[T]he fair treatment of people of all races, cultures and income with respect to development, adoption and implementation of environmental laws, regulations and policies" and defines "fair treatment" to mean that environmental laws, regulations, and policies should not lead to "disproportionate impacts on low income communities and communities of color" ' and that such communities "share equitably in the benefits" from such laws, regulations, and policies. They add that such communities must be allowed to participate "as equal partners in every level of decision-making" (http://www.ejcw.org/About/water%20justice.htm).

[14] Water justice is about building a communal vision for how water is distributed and managed. Water justice will be achieved when low-income communities and communities of color have access to water for drinking, cooking, swimming, fishing, cultural and other uses. It requires alternative water allocation and use systems counteracting the fundamentally flawed system of water use, distribution and planning in California (http://www.ejcw.org/About/water%20justice.htm).

[15] http://www.deltavision.ca.gov/StrategicPlanningProcess/StaffDraft/Comments/Com ment_from_Environmental_Justice_9-2-08.pdf

[16] The lead author for this report was a former graduate student researcher of two of the researchers of this study.

[17] The report examined publicly available regulatory agency data and census data and drew on three focus groups held in the Delta's urban areas of Tracy, Stockton, and Pittsburg, 15 in-depth interviews with Delta community members and individuals who have worked around issues pertaining to environmental and social justice in the Delta. Finally, the report draws from ethnographic observations (Liévanos et al 2009) of Delta Vision-related meetings held in Sacramento and in and around the Delta.

[18] See http://www.water.ca.gov/deltainit/docs/062308SuisunCity.pdf

References

Agyeman J (2002) Constructing environmental (in)justice: Transatlantic tales. *Environmental Politics* 11:31–53

Arreola D (1975) "Locke, California: Persistence and Change in the Cultural Landscape of a Delta Chinatown." Unpublished Master's thesis, CSU Hayward

BDPAC (2007) CALFED Bay-Delta Program: Program Performance Assessment. http://calwater.ca.gov/calfed/oversight/BDPAC/index.html Accessed October 2008

Blue Ribbon Task Force (BRTF) (2008) *Our Vision for the California Delta*. Sacramento: State of California Resources Agency

Carroll P (2006) *Science, Culture, and Modern State Formation*. Berkeley: University of California Press

Chan S (1987) *This Bitter Sweet Soil: the Chinese in California Agriculture, 1860–1910*. Berkeley: University of California Press

Cole L W and Foster S R (2001) *From the Ground Up: Environmental Racism and the Rise of the Environmental Justice Movement*. New York: New York University Press

Community Water Center (2008) http://www.communitywatercenter.org/ Accessed October 2008

Cutter S L (1996) Vulnerability to environmental hazards. *Progress in Human Geography* 20:529–539

Cutter S L, Boruff B and Shirley W L (2003) Social vulnerability to environmental hazards. *Social Science Quarterly* 84:242–261

Delta Protection Commission (2008) http://www.delta.ca.gov/recreation/economic/default.asp Accessed March 2008

Delta Vision (2008) http://deltavision.ca.gov/AboutDeltaVision.shtml Accessed October 2008

Delta Vision Stakeholder Coordination Group (2007) Preliminary recommendations report prepared for the Delta Vision Blue Ribbon Task Force. http://deltavision.ca.gov/StakeholderReports/Stakeholder_Coordination_Group_Preliminary_Report.pdf Accessed October 2008

Desbiens C (2007) "Water all around, you cannot even drink": The scaling of water in James Bay/Eeyou Istchee. *Area* 39:259–267

Egelko (2008) *San Francisco Chronicle*. Water plan can proceed, high court rules. 6 June

Environmental Justice Coalition for Water (EJCW) (2005) Thirsty for justice: A people's blueprint for California water. http://www.ejcw.org Accessed October 2008

Environmental Justice Coalition for Water (EJCW) (2008) Environmental Justice Delta Vision. http://www.deltavision.ca.gov/DeltaVisionVisions.shtml Accessed January 2008

Fielding J and Burningham K (2005) Environmental inequality and flood hazard. *Local Environment* 10:379–395

Gilmore R W (2007) *Golden Gulag: Prisons, Surplus, Crisis, and Opposition in Globalizing California*. Berkeley: University of California Press

Harrison J L (2006) Accidents and invisibilities: Scaled discourse and the naturalization of regulatory neglect in California's pesticide drift conflict. *Political Geography* 25:506–529

Harvey D (1996) *Justice, Nature, and the Geography of Difference*. Malden, MA: Blackwell Publishers

Heikkila T and Gerlak A K (2005) The formation of large-scale collaborative resource institutions: clarifying the role of stakeholders, science and institutions. *Policy Studies Journal* 33:583–612

Herod A J (1991) The production of scale in United States labor relations. *Area* 23(1):82–88

Herod A J (1997) From a geography of labor to labor geography: Labor's spatial fix and the geography of capitalism. *Antipode* 29:1–31

Heynen N (2003) The scalar production of injustice within the urban forest. *Antipode* 35:980–998

Hillman M (2006) Situated justice in environmental decision-making: Lessons from river management in Southeastern Australia. *Geoforum* 37:695–707

Hundley N J (2001) *The Great Thirst: Californians and Water—a History*. Berkeley: University of California Press

Innes J E, Connick S and Booher D (2007) Informality as a planning strategy—collaborative water management in the CALFED Bay–Delta Program. *Journal of the American Planning Association* 73:195–210

Isenberg A C (2006) *Mining California: An Ecological History*. New York: Hill & Wang

Jacobs K L, Luoma S N and Taylor K A (2003) CALFED: An experiment in science and decision-making. *Environment* 45:30–41

Johnson S, Haslam G and Dawson R (1993) *The Great Central Valley: California's Heartland*. Berkeley: University of California Press

Jones P and Macdonald N (2007) Making space for unruly water: Sustainable drainage systems and the disciplining of surface runoff. *Geoforum* 38:534–544

Kaika M (2005) *City of Flows: Modernity, Nature, and the City*. New York: Routledge

Koehler C (1995) California's Bay-Delta Agreement: a model for cooperation. *Rivers* 5:46–51

Kurtz H E (2003) Scale frames and counter-scale frames: Constructing the problem of environmental justice. *Political Geography* 22:887–916

Liévanos R, Davis D, Kim M and Davis C (2009) Third parties no more: Envisioning an environmentally just and sustainable Sacramento-San Joaquin Delta. http://www.ejcw.org Accessed March 2008

Little Hoover Commission (2005) Still imperiled, still important: The Little Hoover Commission's Review of the CALFED Bay-Delta Program. http://www.lhc.ca.gov/lhcdir/listall.html Accessed January 2008

London J, Sze J and Liévanos R (2008) Problems, promise, progress, and perils: Critical reflections on environmental justice policy implementation in California. *UCLA Journal of Environmental Law and Policy* 26(2):255–289

Loo T (2007) Disturbing the peace: Environmental change and the scales of justice on a northern river. *Environmental History* 12:895–919

Lund J, Hanak E, Fleenor W, Howitt R, Mount J and Moyle P (2007) Envisioning futures for the Sacramento-San Joaquin Delta. Public Policy Institute of California. http://www.ppic.org Accessed August 2008

McPhee J (1990) *The Control of Nature*. New York: Farrar, Straus, and Giroux

McWilliams C (1999 [1935]) *Factories in the Field: The Story of Migratory Farm Labor in California*. Berkeley: University of California Press

Mitchell M D (1996) The Sacramento–San Joaquin Delta, California: Initial transformation into a water supply and conveyance node, 1900-1955. *Journal of the West* 35:44–53

Nash L (2006) *Inescapable Ecologies: A History of Environment, Disease, and Knowledge.* Berkeley and Los Angeles: University of California Press

Openshaw S (1983) *The Modifiable Areal Unit Problem.* Norwich: Geo Book

Pellow D N (2007) *Resisting Global Toxics: Transnational Movements for Environmental Justice.* Cambridge, MA: MIT Press

Pisani D (1984) *From the Family Farm to Agribusiness: The Irrigation Crusade in California and the West, 1850–1931.* Berkeley: University of California Press

Raley B W (2005) Testimony to the Little Hoover Commission concerning the CALFED Bay-Delta program governance, 25 August. http://www.lhc.ca.gov/lhcdir/calfed/ Accessed October 2008

Robbins P (2007) *Lawn People: How Grasses, Weeds, and Chemicals Make Us Who We Are.* Philadelphia: Temple University Press

Schlosberg D (2004) Reconceiving environmental justice: Global movements and political theories. *Environmental Politics* 13:517–540

Schlosberg D (2007) *Defining Environmental Justice: Theories, Movements, and Nature.* Oxford: Oxford University Press

Shilling F, London J and Liévanos R S (forthcoming A) Third parties and a process of elimination: Environmental justice within and beyond CALFED. Invited submission for special issue on CALFED. *Environmental Science and Policy*

Shilling F, White A and Lippert L (forthcoming B) Contaminated fish consumption in California's Central Valley Delta. *Environmental Health Research*

Silver E, Kaslow J, Lee D, Sun L, May L T, Weiss E and Ujihara A (2007) Fish contamination and advisory awareness among low-income women in California's Sacramento–San Joaquin Delta. *Environmental Research* 104:410–419

Smith N (1984) *Uneven Development: Nature, Capital and the Production of Space.* Oxford: Blackwell

State Water Project (2008) History of water development and the SWP. http://www. publicaffairs.water.ca.gov/swp/history_swp.cfm#cvp Accessed October 2008

Swyngedouw E (1996) The city as hybrid—on nature, society and cyborg urbanism. *Capitalism, Nature, Socialism* 71(25):65–80

Swyngedouw E (1999) Modernity and hybridity: Nature, regeneracionismo, and the production of the Spanish waterscape, 1890–1930. *Annals of the Association of American Geographers* 89:443–465

Swyngedouw E (2004) *Social Power and the Urbanization of Water: Flows of Power.* Oxford: Oxford University Press

Swyngedouw E and Heynen N C (2003) Urban political ecology, justice, and the politics of scale. *Antipode* 35:898–918

Sze J and London J (2008) Environmental justice at the crossroads. *Sociology Compass* 2(4):1331–1354. http://dx.doi.org/10.1111/j.1751-9020.2008.00131.x Accessed June 2009

Thompson J (1957) "The settlement geography of the Sacramento–San Joaquin Delta, California." Unpublished PhD dissertation, Stanford University

Towers G (2000) Applying the political geography of scale: Grassroots strategies and environmental justice. *The Professional Geographer* 52:23–36

US Bureau of Reclamation (2007) Shasta Lake water resources investigation: Plan formulation report. http://www.usbr.gov/mp/slwri/docs/plan_form_rpt_12-2007.pdf Accessed October 2008

Villarejo D, Lighthall D, Williams D, Souter A, Mines R, Bade B, Samuels S and McCurdy S (2000) Suffering in silence: A report on the health of California's agricultural workers. California Institute for Rural Studies. Woodland Hills, CA: California Endowment

White D and Wilbert C (2009) Introduction: Inhabiting technonatural time-spaces. In D White and C Wilbert (eds) *Technonatures: Environments, Technologies, Spaces and Places in the Twenty-first Century* (pp 1–30). Ontario: Wilfrid Laurier University Press

Williams R W (1999) Environmental injustice in America and its politics of scale. *Political Geography* 18:49–73

Wolff J (2003) *Delta Primer: A Field Guide to the California Delta*. San Francisco: William Stout Publishers

Worster D (1985) *Rivers of Empire: Water, Aridity, and the Growth of the American West*. Toronto: Pantheon Books

Index

Printed and bound by CPI Group (UK) Ltd, Croydon, CR0 4YY

27/10/2024

14580369-0003